IEE TELECOMMUNICATIONS SERIES 31

Series Editors: Professor J. E. Flood
                  Professor C. J. Hughes
                  Professor J. D. Parsons

Data communications and networks 3

Other volumes in this series:

Data communications and networks 3

Edited by
R.L.Brewster

The Institution of Electrical Engineers

Published by: The Institution of Electrical Engineers, London,
United Kingdom

© 1994: The Institution of Electrical Engineers

British Library Cataloguing in Publication Data

A CIP catalogue record for this book
is available from the British Library

ISBN 0 85296 804 3

Printed in England by Redwood Books, Trowbridge, Wiltshire

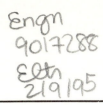
Contents

Preface

The requirement for data communication is closely linked to the invention and development of the digital computer. The first commercially viable computers emerged during the late 1950s and were relatively simple devices, accessed by a single terminal directly associated with the computer hardware. However, the data handling capacity of computers was soon realised and in the early 1960s an airline seat reservation scheme was introduced which required access to a central computer from terminals situated remotely from the computer mainframe. Thus data transmission was born. In the ensuing 30 years there has been an enormous growth in computing power and facilities and in data processing applications. The provision of data communication facilities has, therefore, likewise increased beyond the wildest imaginations of those of us involved with data transmission at its inception in the 1960s.

In the earliest days of data transmission, the only readily available medium with easy access to almost any location in the world was the telephone network. Although not designed to convey digital signals, its very ubiquitousness made it a prime candidate for the transmission of data. Despite its limitations, it served well as the almost exclusive carrier of data signals for a decade of more. In order to interface with the analogue environment of the telephone network, modems were designed to convert the digital signals into signals with a spectral content more like that of telephone speech signals. The first modems appeared in public service in the UK in 1965 and offed data throughput rates of 1200 bits/s half-duplex or 200 bits/s full-duplex over the switched telephone network. Since that date more and more sophisticated modems have been developed using complex modulation schemes and incorporating adaptive equalisation. These have extended the data rates available over the telephone network up to a rate of 19200 bits/s. This seems to be an upper limit and even this requires specially selected lines.

The inadequacy of the telephone network for many data applications was soon realised and dedicated data networks were developed for specific applications. A range of local area networks (LANs) using ring and bus topologies were developed for networks confined to a factory complex or campus. Larger private 'wide area networks' (WANs) were developed for national, and even international, use. Public data networks employing packet switching were introduced for specifically data communication purposes.

Alongside the development of specifically data networks, an interesting development was taking place in telephony. The transmission of telephone signals was gradually being changed from an analogue to a digital mode of transmission using pulse-code-modulation. This meant that trunk telephone transmission was now to be carried on circuits with a digital capability based on multiples of 64 kbits/s, the basic pcm rate per telephone channel. 30 channel pcm uses a line rate of 2 Mbits/s. With the introduction of digitally transmitted telephony, it became an obvious progression to introduce digital switching into the telephone network. This has now become a reality and has led to the introduction of an integrated services digital network (ISDN) which handles all signals, telephony or data, on a single network. However, in the meantime, rapid development in optical fibre technology has led to large capacity 'wide-band' transmission facilities becoming available which can provide data capacities well in excess of 1 Gbits/s. The next step was therefore the introduction of a broadband ISDN (B-ISDN) network which is expected to be widely available for general use by the turn of the century.

The growth in the use of large data networks and in the variety of tasks that the host computers are being asked to perform means that there was an urgent need for standardisation of network interfaces and protocols. The concept of 'Open Systems Interconnection' was therefore introduced which has led to the production of a range of recommendations and specifications for interconnection standards. The purpose of this book is to survey the current state-of-the-art in data communications and networks and to indicate the direction in which networks are likely to be developing in the future. Details of standards and interfaces have been included because, without these, no reasonable network strategy would be possible.

The book is based on notes prepared for the Institution of Electrical Engineers Vacation School on 'Data Communications and Networks' held at Aston University in July 1993. It supersedes the earlier editions of the book, based on the notes of the same schools held in September 1985 and September 1989. Much has happened in the intervening years, especially with regard to the development of ISDN and B-ISDN and in the development of standards for the higher level protocols of the OSI model. The previous books have enjoyed wide popularity but now appear very dated. It is hoped that this new edition, which has been extensively rewritten, will be as popular as its predecessors.

The 1993 Vacation School was video-recorded to produce a distance learning package, thus making the course available to a much wider audience. This book forms a complementary part of the package, which is available from the IEE.

I should like to thank the authors of the various chapters for their valuable contribution to the book. The preparation of the manuscripts has made a significant demand on the valuable time of many busy people and without their dedicated help this book would not have been possible. Thanks are also due to those unnamed, but nevertheless very important people, who have advised, typed, proof-read and helped in various ways with the production of the book.

Ron Brewster
Aston, July 1994

List of contributors

Chapter 1
Dr Ron Brewster
Department of Electronic Engineering
    and Applied Physics
Aston University
Birmingham

Chapter 2
John Ash and Don Richards
GPT Limited
Coventry

Chapter 3
Prof. Fred Halsall
Department of Electrical and Electronic
    Engineering
University of Wales
Swansea

Chapter 4
John L. Moughton
Cray Communications
Swindon, Wilts.

Chapter 5
Ron Haslam
BT Laboratories
Martlesham Heath
Ipswich

Chapter 6
Ian Gallagher
BT Laboratories
Martlesham Heath
Ipswich

Chapters 7 and 8
Mrs A Gill Waters
Computing Laboratory
University of Kent at Canterbury
Canterbury

Chapters 9 and 10
Keith Caves
Network Technology Division
BNR Europe Limited
Harlow, Essex

Chapter 11
Ian Harris
Vodata Limited
Newbury, Berks.

Chapter 12
John P. Chambers
Formerly at BBC Reasearch Department
Now at National Physical Laboratory
Teddington, Middlesex

Chapter 13
Ron Bell
Information Technology Services Ltd.
Galleywood
Chelmsford, Essex

Chapter 1

Introduction and overview

Ron Brewster

The requirement for data communication is closely linked to the invention and development of the digital computer. The first commercially viable computers emerged during the late 1950s and were relatively simple devices, accessed by a single terminal directly associated with the computer hardware. However, the data handling capacity of computers was soon realised and in the early 1960s an airline seat reservation scheme was introduced which required access to a central computer from terminals situated remotely from the computer mainframe. Thus data transmission was born. In the ensuing 30 years there has been an enormous growth in computing power and facilities and in data processing applications. The provision of data communication facilities has, therefore, likewise increased beyond the wildest imaginations of those of us involved with data transmission at its inception in the 1960s.

In the earliest days of data transmission, the only readily available medium with easy access to almost any location in the world was the telephone network. Although not designed to convey digital signals, its very ubiquitousness made it a prime candidate for the transmission of data. Despite its limitations, it served well as the almost exclusive carrier of data signals for a decade or more. In order to interface with the analogue environment of the telephone network, modems were designed to convert the digital signals into signals with a spectral content more like that of telephone speech signals. The first modems appeared in public service in the UK in 1965 and offered data throughput rates of 1200 bit/s half-duplex or 200 bit/s full-duplex over the switched telephone network. Since that date more sophisticated modems have been developed using complex modulation schemes and incorporating adaptive equalisation. These have extended the data rates available over the telephone network to a rate of 9600 bit/s. This seems to be an upper limit and even this requires specially selected lines.

The inadequacy of the telephone network for many data applications was soon realised and dedicated data networks were developed for specific applications. A range of local area networks (LANs) using ring and bus topologies were developed for networks confined to a factory complex or campus. Larger private 'wide area networks' (WANs) were developed for national, and even international, use. Public data networks employing packet switching were introduced for specifically data communication purposes.

Alongside the development of specifically data networks, an interesting development was taking place in telephony. The transmission of telephone signals was gradually being changed from an analogue to a digital mode of transmission using pulse-code-modulation (pcm). This meant that trunk telephone transmission was now to be carried out on circuits with a digital capability based on multiples of 64 kbit/s, the basic pcm rate per telephone channel. 30 channel pcm uses a line rate of 2Mbit/s.

With the introduction of digitally transmitted telephony, it became an obvious progression to introduce digital switching into the telephone network. As this became a reality, the next logical step was the introduction of an integrated services digital network (ISDN) which handles all signals, telephony or data, on a single network. Present expectations are for ISDN to be in common use at least by the turn of the century, with developments towards the general provision of broadband ISDN (B-ISDN) well under way by this time.

The growth in the use of large data networks and in the variety of tasks that the host computers are being asked to perform means that there is an urgent need for standardisation of network interfaces and protocols. Much has happened recently in the area of strategies for 'Open Systems Interconnection' and in the production of recommendations and specifications for interconnection standards. The purpose of this book is to survey the current state-of-the-art in data communications and networks and to indicate the direction in which the networks are likely to be developing in the future. Details of standards and interfaces have been included because, without these, no reasonable network strategy would be possible.

1.1 Back to basics

Data is the very essence of computing; every computer operation involves the manipulation of data in one form or another. To carry out meaningful operations, the computer has first to be supplied with the appropriate data. Data can be fed into the computer in a number of ways. Familiar methods are through keyboards, card and tape readers, disk drives and optical character readers. Other more specialist devices also exist, such as measurement transducers and proximity sensors.

It is pointless manipulating data in a computer if it is not subsequently to be made available for further use. Typical data output devices are visual display units, printers, control transducers and alarms. Processed data may also be stored on cards, tape or disk so that it can be used again at a later date. In large computer networks data is often exchanged directly between computers within the network. The efficient transmission of data to and from data terminal equipment, sometimes referred to as computer peripherals, and between computers themselves, is therefore a vitally important function.

When computers first came into general use in the early 1950s they were self-contained units with terminals which were either integral with the processor or were installed adjacent to the main equipment. There was thus no problem of interconnection; the units were simply connected together using multi-core cable with sufficient wires of adequate quality to carry separately all the necessary data and control signals. Very quickly, however, it became desirable to separate the terminal from the central processor at distances where it was no longer economical to use multicore cables. Instead, a cable consisting of a single wire pair was used and techniques for combining the data and control signals into a single data stream were devised.

The next development was the need to operate from terminals from remote locations, requiring the transmission of data over lines passing outside the site at which the computer mainframe was situated. This need arose through the growth of centralised control systems, such as those operated by Gas Boards and Water Authorities, and the demand for Computer Bureau facilities.

At the same time, in the UK, all communication facilities outside the private site were still the sole prerogative of British Telecom (then the Post Office). The provision of data transmission facilities offered a new challenge; until this time the vast majority of communication traffic had been telephony, though for many years a relatively small amount of telegraphy had also been carried. Because of its ready availability, it was obviously desirable to use, if at all possible, the public telephone network to carry the new data traffic from remote terminals to the centralised computer facility. The lines, however, had not been designed with digital signals in mind, and their characteristics thus appeared to be inappropriate for this application. In order to match the digital signals to the line characteristics, MODEMS were developed to interface the digital equipment to the telephone network. These modems were provided by the Post Office as part of the data transmission facility offered under the commercial name of DATEL services.

Because of limitations set by noise and channel bandwidth, the maximum data rate obtainable over the public switched telephone network is about 4.8 kbit/s, although rates of 9.6 kbit/s are available over special-quality private leased telephone circuits. More recently, it has become common practice to transmit speech signals in the junction and trunk telephone network using pulse-code-modulation (pcm). In pcm the speech signal is sampled and quantised and converted into a digital signal at a transmission rate of 64 kbit/s. This facility may be utilised for the transmission of data, the service being offered in the UK by British Telecom under the title KILOSTREAM. In this way, a data rate of 64 kbit/s can be achieved over the equivalent of one single telephone speech circuit. Alternatively, the equivalent of 30 pcm speech channels can be used as a single data transmission facility operating at 2.048 Mbit/s, the service being offered under the title MEGASTREAM. With the growth of the digital speech network and the introduction of digital switching techniques, an 'Integrated Services Digital Network' (ISDN) is now emerging which will carry both data and digitally-encoded speech signals without discrimination. The provision of such a service will ultimately make the DATEL services, together with their expensive modems, obsolete.

The concept of the DATEL services is to provide a 'transparent' data transmission facility such that the provider of the carrier service has no interest in either the format or the significance of the transmitted data stream. In order to allow terminals to be readily connected together via the network, a standard equipment interface was established by CCITT. This interface is widely known as the V24 interface. A similar (though not identical) interface is in use in the USA and is known as the RS232 interface. The interface not only facilitates the actual transfer of data between the data terminal equipment (DTE) and the line terminating equipment (LTE), but also provides certain control signals which allow some interaction between the terminal and the network.

With the advance of computer technology and the widespread introduction of the microprocessor, terminal equipment has become more sophisticated and distributed computing power has become commonplace. Data networks have therefore been set up to allow computers and data terminals to communicate together as required by the user. Most networks of this type are set up within the confines of a factory site, university campus or office block. These networks do not, therefore, make use of public network transmission facilities. They operate on wideband circuits, data being multiplexed and assigned using a suitable network protocol. Such networks are known as 'Local Area Networks' or LANs.

Many large organisations now operate data networks on a regional, national, or even international basis provided by the interconnection of LANs to form Metropolitan Area Networks (MANs) or Wide Area Networks (WANs). This has created an ever-increasing demand for data transmission facilities in the public domain. Initially this was provided by the use of modems transmitting analogue-type signals over the PSTN to represent the digital data symbols. This technology is still in use, although nowadays much more use is made of the digital capability within the PSTN in the form of services such as the British Telecom KILOSTREAM (64 kbit/s) and MEGASTREAM (2 Mbit/s) services. There are, at present, considerable limitations on these services, since access to the digital service is only available at the local exchange access to the trunk telephone network.

In the conventional "circuit-switched" operation of the telephone network, a circuit is made available to a user throughout the whole duration of a communication session whether or not he is actually transmitting information. This can lead to gross under-utilisation of the available transmission capability when used for many data applications. To overcome this, packet-switched networks were introduced. This technique allows a transmission resource to be shared among a number of users on a statistical basis and therefore results in much more efficient use of the resource. However, the technique can lead to some packet delay, which can, in some circumstances, vary from packet to packet. Packet switching is therefore not appropriate for digital speech transmission, unless quite complex priority techniques are introduced. Packet switching has enjoyed varied popularity and has not been so widely used in the public domain in the UK as in other parts of Europe and particularly in the United States, where it was introduced fairly early on in the use of digital communication for computer applications.

As ISDN and, later, broadband ISDN (B-ISDN) become more widely available, it will, no doubt, form the basis for almost all wide area network operation. The initial ISDN service is based on the general provision of 64 kbit/s capacity channels. At the time of writing ISDN is only available on a limited basis because not all exchanges have been converted yet to digital switching technology. However, the conversion process is proceeding at such a rate that most exchanges, at least in the UK, are by now completely digital. Pilot schemes have been in operation in the United States and parts of Europe, notably the UK, France and Germany, since around the mid 1980s. Most of these pilot schemes are not fully to the CCITT ISDN recommendations because they were introduced into service before the CCITT standards were definitively established. Most network operators world-wide have a programme of modernisation which will make digital switching more or less universal by the end of the century. In the meantime, the availability of ISDN is necessarily restricted to those users served by local exchanges which are fully connected end-to-end using digital technology. It is also necessary to provide digital local access between the user and the local exchange. Because of the enormous investment in local line plant, it is not possible to replace, at least in the short term, all the circuits used to connect the user to the local exchange. A typical local network is shown in Fig. 1.1.

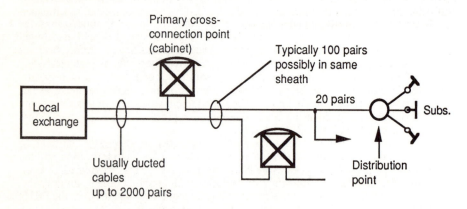

Fig. 1.1 *Typical local network*

The local network has been provided with the transmission of analogue speech signals of limited bandwidth in mind. A connection from user to exchange may involve several cable sections, the wire pairs in each section being of different gauges and grouped together in cables of various sizes. The much greater bandwidth required for digital transmission at 64 kbit/s or higher and the increased effect of crosstalk at these frequencies means that considerable ingenuity has to be used to obtain satisfactory digital transmission over the local lines. The techniques that have been adopted are discussed in a later chapter.

Even when ISDN becomes widely available, the main uses will probably not be very different from those in current vogue, although much more efficient utilisation and greater flexibility in control will be available to the network user. ISDN can also provide a 2 Mbit/s primary-rate service to business users to replace the MEGASTREAM service widely used at the current time. This will provide greater flexibility in the use of the primary-rate facility than is available with MEGASTREAM through the use of enhanced signalling procedures provided by Stored Program Control (SPC).

The initial ISDN service is sometimes nowadays referred to as N-ISDN, standing for narrowband ISDN, to distinguish it from broadband B-ISDN. The broadband ISDN will be based on optical fibre transmission and will provide much wider access to bit-rates of the order of hundreds of Mbit/s. Eventually, this is likely to extend to domestic users. This capability will make digital video services such as cable-distributed TV and videophone generally available, as well as many other information services based on video techniques. Broadband ISDN is likely to have had a very considerable impact on the lives of almost every living person by the end of the first decade of the 21st century. The possibilities for the use of the available transmission capability are only just beginning to be explored in any depth. One thing is certain; new services as yet unconceived of are likely to become available with surprising rapidity once the network begins to penetrate into the environment. The popularity of cellular telephony and facsimile transmission in the last couple of years has far exceeded any expectations. It would be surprising if B-ISDN does not enjoy the same enthusiastic reception.

The development of data networks has led to the concept of Open Systems Interconnection (OSI). The purpose of OSI is to enable a wide variety of data terminals from different manufacturers to be interconnected freely over common data transmission facilities. This has meant the development of specifications and network protocols which enable such operation to take place. These specifications are based on a 7-layer protocol structure proposed by the International Standards Organisation (ISO) and known as the ISO reference model for OSI. The structure and significance of this model is discussed in a later chapter.

Chapter 2

Data over analogue systems

John Ash and Don Richards

2.1   Introduction

The convergence between computing and telecommunications is now an accepted fact
of modern communications networks. Today's widespread application of computers
for billing, information storage and retrieval, banking, word-processing, electronic
mail, the 'home office' and many others could not have taken place without the ability
to transfer information between computers and terminals over long distances. Since its
early beginning in the 1960s, improvements in the technology have enabled transfer of
data to take place at higher and higher rates and with greater economy.

Over the past decade, considerable advances have been made in the introduction of
digital transmission into the trunk and junction network in the UK but , so far, its
penetration into the local network has been limited. As far as the telephone user is
concerned the only available widespread network for carrying data has been the
existing analogue telephone network. In spite of the introduction of special digital data
networks (such as packet switching and circuit switched digital services), this is likely
to remain the most readily accessible widespread medium for some time to come.

Unfortunately, the telephone network was designed with speech signals only in
mind and hence little regard was paid to the problems of phase and amplitude variation
with frequency, which severely limits the performance for data transmission but has
little or no significance for speech. In addition, the presence of ac-coupled
transmission bridges and the use of channel shaping filters in the network excludes a
dc path and effectively limits the useful frequency band to 300Hz-3400Hz. The change
from a frequency division multiplex system to an all digital network has helped both
the bandwidth availability and the group delay distortion. However, such a medium is
unsuited to the direct transmission of digital signals, whose characteristics consist of
rectangular pulses, producing spectral components from dc to several kilohertz. An
example of the sort of problem which might be encountered is illustrated for a worst
case PSTN connection in Fig. 2.1(a) and (b). The translation of digital signals into a
form more suited to voiceband transmission is performed by a modem.

The primary purpose of a modem is to provide the means for transmitting and
receiving data signals in an analogue environment, but it is important to recognise that
other functions are necessary to provide a 'black box' approach to network
connectivity (Fig. 2.2). A more detailed arrangement of the modulator and
demodulator functions for a typical modem is illustrated in Fig. 2.3 and discussed later
in the text.

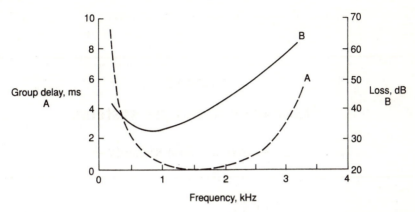

Fig. 2.1(a) *Representative worst case PSTN connection for FDM systems*

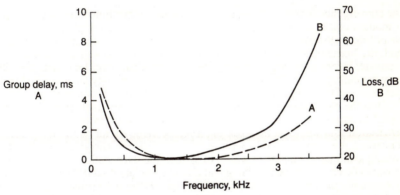

Fig. 2.1(b) *Worst case PSTN connection for digital systems*

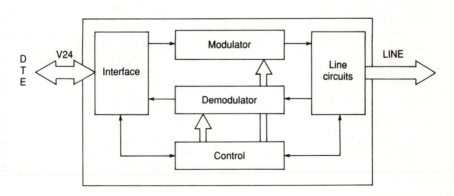

Fig. 2.2 *The modem as a 'black box'*

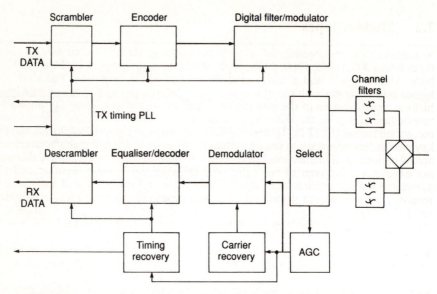

Fig. 2.3 *A typical modulation/demodulation process for a V22bis modem*

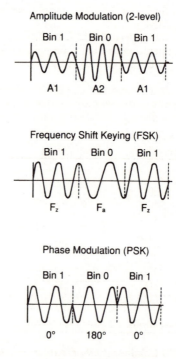

Fig. 2.4 *Representation of amplitude, frequency and phase modulation*

2.2 Modem types

Voiceband modems generally fall into the categories governed by their user bit rates, these being 300, (600), 1200, 2400, 4800, 9600 and more recently 14400 and 19200 bit/s. Although 600 bit/s is rarely used these days, it will be noticed that all of the higher bit rates are a multiple of this basic rate. In fact, this sequence is virtually an historic representation of the evolution of modems, as the technology for higher bit rates came along. Each of the speeds has one or more associated international recommendations (CCITT) [1], defining the modulation and control employed, and the key parameters necessary to ensure compatibility across different manufacturers and across national and international boundaries. A summary of this grouping, although not exhaustive, is given in the Table 2.1, showing the V-series number and the associated analogue channel line rate and method of separation (sep) of transmit and receive for 2-wire duplex working (FDM = frequency division multiplex, i.e. filtering and EC = echo cancelling). In addition to the V-series recommendations listed there, it is perhaps worth mentioning other specifications significant at this stage.

V24   Defines the functions of the various data control and interchange circuits, which operate across the modem-terminal interface.

V25   Defines the operation of automatic answering and the parallel interface for automatic calling modems.

V25bis   Defines the operation of automatic answering and the serial interface for automatic calling modems

V42   Defines the operation of two alternative error correction protocols - LAP M and MNP 4 [2].

V42bis   Defines the operation of a data compression protocol for asynchronous data running under the error correcting protocol LAP M.

MNP 5.[2] Proprietary data compression algorithm for asynchronous data running under the error correction alternative MNP 4 from recommendation V42. This is a de-facto standard not recognised by CCITT.

CCITT Rec.	Line speed(bit/s)	Modulation	Line Format (sep)
V21	0 - 300	FSK	2-wire duplex (FDM)
V22	600 - 1200	DPSK	2-wire duplex (FDM)
V22bis	1200 - 2400	QAM 16 state	2-wire duplex (FDM)
V23	1200/75 - 75/1200	FSK	2-wire duplex (FDM) or half duplex 1200/1200
V26	1200 - 2400	DPSK	4-wire duplex
V26bis	1200 - 2400	DPSK	2-wire half duplex
V26ter	1200 - 2400	DPSK	2-wire duplex (EC)
V27	2400 - 4800	DPSK	2-wire half duplex
V29	4800 - 9600	QAM 16 state	2-wire half duplex
V32, bis	4800 - 14400	QAM + trellis	2-wire duplex (EC)
V33	4800 - 14400	QAM + trellis	4-wire duplex

Table 2.1 *Common CCITT modem recommendations*

The term 'bit rate' refers to the rate at which each binary digit (bit), having the value '1' or '0', is transferred across the interface. This should not be confused with the term 'baud' (in symbols per second), which is used to express the signalling rate over the analogue channel itself.

A further area of delineation in modem terminology is the distinction between SYNCHRONOUS and ASYNCHRONOUS modems. An asynchronous modem conveys no timing information across the data channel and therefore can only accept start/stop or truly random 1/0 transitions. In the case of synchronous operation, the modem accepts and transmits timing information along with the data, such that each element of the data is synchronised with the modem clock. Not only is this latter more efficient, it is a prerequisite of the more complex modulation techniques employed at bit rates of 1200bit/s and above. Where an essentially synchronous modem is required to operate in an asynchronous application the start/stop data is synchronised to the modem clock before transmission synchronously to the receiving modem, where it is reconverted to asynchronous format. This technique is used from V22 upwards and is covered by CCITT recommendation V14.

2.3 Modulation methods

2.3.1 Line coding (baseband) techniques

The effects of group delay and amplitude variations encountered on telephone circuits cause severe distortion of rectangular data pulses, to the extent that detection and restoration of the original pulse train can become very difficult.

For relatively short distances, for example over copper in the local network, so-called 'baseband' or 'line coding' techniques [3] can be used to give reliable pulse transmission. This usually involves modifying the original unipolar pulse train from the DTE (data terminating equipment) by an encoding process which converts it into a form having more useful transmission properties. In particular, the coding would seek to eliminate the dc and lower frequency components of the original train to maximise the number of element transitions to assist timing recovery. In many cases advantage is also taken of shaping the pulses (for example, concentrating the signal energy towards the higher frequencies can compensate for increasing line length attenuation), by either pre-distorting the pulse with suitable filters or by operating directly on its spectral distribution by the encoding process itself, as in the case of the once popular 'WAL 2' code.

Baseband techniques are used extensively in the comparatively low cost short haul modems, for communicating across cities, where the maximum distances are in the range 10 to 20km, and where a conventional modem would be prohibitively expensive especially at the higher bit rates. For greater distances, extending beyond the local network and into the junction and trunk network, transmission becomes more difficult due to the line distortion effects of phase and frequency modulation in the network. Because of this, it is necessary to translate the original digital baseband signal into a useable portion of the telephone channel, by causing it to modulate a carrier frequency, suitably centred conveniently within the voiceband range. The resulting line signal, 's(t)', can be expressed by

$$s(t) = A(t)\left\{ 2\int_0^t f(T)d(T) + \phi(t) \right\}$$

Thus we may choose to cause the incoming data signal to vary one (or more) of the amplitude 'A', the frequency 'f', or the phase 'ø', in the modulation process. A pictorial illustration of these approaches is shown in Fig. 2.4 (p.8).

2.3.2   Frequency modulation

One of the simplest modulation methods is Frequency Shift Keying (FSK), where each binary '1' or '0' is represented by one of a pair of tones; each of which corresponds to one or other of the binary input digits. For very low bit rates, we may imagine the two tones to approximate to two corresponding sets of spectral lines. As the bit rate increases, the spectrum expands on both sides of the chosen tone frequency and the practical limit is achieved at about 1200bit/s (or more correctly 1200 baud) where the bandwidth containing the whole spectrum has reached that provided by the telephone channel.

For a given bit rate, FSK signals occupy a wider bandwidth than other modulation schemes, but it does have the advantage of a constant power level and a better signal to noise ratio (SNR) performance. Both generation and detection of FSK signals are relatively simple, which helps to explain the early choice of FSK for modems up to 1200bit/s. Today, modern technology has leant itself to low price realisation of both V21 and V23 modems and their Bell (USA) equivalents - often now all combined onto a mixed technology single chip.

2.3.3   Amplitude modulation

In Fig. 2.5, the carrier frequency 'f_c', is directly modulated by the baseband data signal, 'f_m'. The resultant amplitude modulation (AM) signal has both an upper and a lower sideband, each of which contains identical information, together with the original carrier component. Because the double sideband signal is naturally limited by the modulation rate to exclude a dc component and the higher frequencies, it is suitable for transmission in a 300 - 3400Hz telephone bandwidth system. A good choice of modulation rate and carrier frequency can give a workable system, and in fact some modems in the early 1970s used a form of amplitude modulation, but they were later replaced by more bandwidth-efficient modulation schemes.

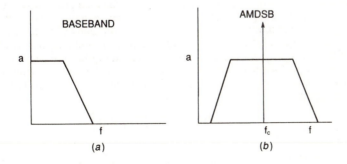

Fig. 2.5 *Amplitude modulation spectra*

2.3.4 Phase modulation

A more commonly used modulation scheme is Phase Shift Keying (PSK), in which the data elements are used to change the 'phase' of the carrier 'f_c' by a fixed amount (say 0° or 180°), corresponding to the binary data '1' or '0'. At the receiver, a derived reference carrier is used to decode the absolute phase directly. The signal spectrum generated is similar in shape to the amplitude modulation spectrum. Although this so-called 'coherent' detection offers a theoretically optimum performance, it is easily perturbed in practice, leading to problems of phase ambiguity in the detection process. It is more usual, therefore, for each data element to be transmitted as a 'phase change' relative to that of the preceding data element. At the receiver, decoding is then accomplished by determining the relative phases of consecutive samples of the line signal. This Differential Phase Shift Keying (DPSK) is most commonly adopted even though it does result in an overall reduction in the signal to noise performance compared with coherent detection. Using a DPSK system avoids the necessity of having an absolute carrier reference and the demodulating carrier is now usually phase locked to a long term average of the incoming signal, which also helps to recover the lost signal to noise performance.

2.4 Multilevel coding

According to Shannon's classic formula [4], the information carrying capacity 'C'(bit/s) for a continuous channel of bandwidth 'B'(Hz) is given by the expression below and can be evaluated for white gaussian noise with signal-to-noise ratio (S) of say 30dB and a telephone bandwidth of 2700Hz.

$$C = B.\text{Log}\ (1+S)$$

(giving C = 27000bit/s for an ideal distortionless channel)

However, if the same values are applied to Nyquist's criterion [5] for eliminating intersymbol interference between successive elements in an ideal bandlimited channel, then the maximum theoretical bit rate is given by

$$b = 2B\ \text{bit/s}$$

(giving b = 5400bit/s)

This figure is substantially lower than Shannon's channel capacity and the double sideband spectrum form of DPSK would indicate a further reduction to 2700bit/s. In practice, the data rate can be taken beyond the Nyquist criterion limit, by increasing the number of states the carrier can adopt. For example, if two data bits at a time are encoded, then the four resultant dibit (two bit) combinations '00', '01', '10' and '11' would need to be represented by one of the four possible states of the carrier. But because each dibit pair can now be transmitted on alternate time boundaries, the modulation rate becomes half of the incoming data stream bit rate. Therefore, provided the line rate is below 2700bit/s(baud), the Nyquist criterion can be upheld. In general, if n bits at a time are encoded, we require 2n possible carrier states or in Nyquist terms

$$b = 2nB\ \text{bit/s}$$

The penalty that must be paid for this increase in bandwidth utilisation is that not only will the modem detector need to be more complicated, but there will also be a degradation in the SNR performance in going from 2 to 2n states. For example, comparing 2-state 2400bit/s operation with 4-state 4800bit/s in the same bandwidth results in a reduction of 6dB in the SNR margin. In spite of this kind of limitation, rates of 9600bit/s are now quite normal over the telephone circuits and reasonable performance has been achieved up to 19200bit/s. To achieve these results the line rates are in the range 2400 to 3200 baud with only slightly more complicated encoding patterns.

2.4.1 Multilevel Differential Phase Shift Keying

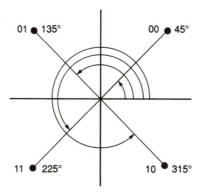

Fig. 2.6 *4-phase DPSK, corresponding to CCITT Recommendation V26 (altB)*

An example of multilevel coding for a 4-state DPSK scheme is illustrated in Fig. 2.6, in which two binary elements are coded for each phase state. This scheme is employed in CCITT Recommendation V26. The corresponding line signalling rate is 1200 baud and with V26 alternative B there is no need for a scrambler because the advancement of the phase at every boundary by at least 45° ensures that receiver synchronism is maintained.

2.4.2 Quadrature Amplitude Modulation

In principle, one can extend the DPSK approach still further to 16 phases (equivalent to 9600bit/s) but, by including some amplitude modulation component, a more efficient use of the carrier modulation possibilities is achieved, giving the so-called Quadrature Amplitude Modulation (QAM) scheme. However, for each such increase in the possible modulation combinations, the margin against noise and phase impairments in the telephone channel becomes progressively less. Savings in design complexity can be made by exploiting the fact that a DPSK signal can be considered as two separate baseband signals, each transmitted independently on a single carrier having the same frequency but with a 90° phase difference. These In-phase(I) and Quadrature(Q) components of the same carrier produce a composite signal that can be decoded in a similar way by generating I and Q recovered carriers for the demodulation process.

In the simplified arrangement of Fig. 2.7, an incoming bit stream is encoded onto I and Q streams representing the discrete phase and amplitude values or 'constellation points' of the particular modulation plan. The resulting composite QAM signal constellation is illustrated in Figs. 2.8 and 2.9 for two different types of 16 point constellations. By inspection it can be seen how each can be translated into one of four values on each of the I and Q axes. The V29 star-shaped constellation has fewer phases and is an older recommendation; the square format of the later V22/V32 is designed specifically round the I and Q modulation and demodulation principle where the extra number of phases in the constellation is not reflected in any increase in the complexity of the system.

The tolerance to error is determined by the proximity of one point in the constellation to its nearest neighbour and various proposals have been suggested to optimise the geometric structure with regard to maximising the performance for gaussian noise, phase jitter, phase and amplitude hits, differential encoding, peak and mean power, ease of implementation and so on. Once again, the penalty which must be paid for this increase in bandwidth efficiency manifests itself as an increase in modem detection complexity. Perturbations in the received line signal will cause errors in both the I and Q channels, i.e. each encoded channel will contain cross interference from its quadrature channel. Thus an extra equaliser is usually included in addition to the normal telephone equaliser to remove the unwanted effects. Nowadays, such an arrangement is well within the scope of modern digital signal processors (DSP), whereas ten years ago the implementation would have been prohibitively expensive.

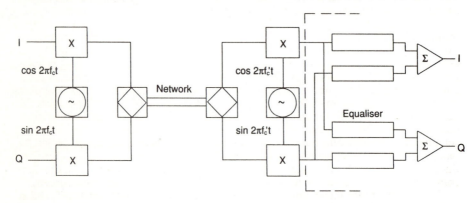

Fig. 2.7 *A Simplified QAM system*

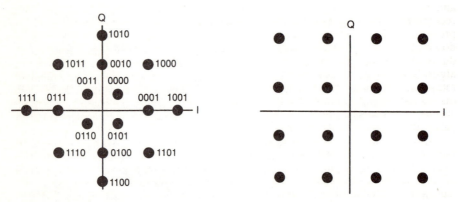

Fig. 2.8   *V29 signal space diagram (16 point)*   **Fig. 2.9**   *V22/V32 signal space diagram (16 point)*

2.5 Line signal coding

Although modems have been produced employing large numbers of points in their signal constellations, there results a need for increasingly complex detection schemes, capable of achieving the expected bit-error-rate performance under all conditions of line and circuit impairments. Various coding schemes have been developed to improve the overall error-rate in data transmission systems. This mostly involves the inclusion of redundant coding bits in the data encoding process within the modem transmitter, to permit the use of error enhancement techniques in the receiver. That is to say, the current value of the transmitted data symbol can be made dependent on the state of previously transmitted data symbols, either immediately or in some convoluted manner. But, because any extra bits must increase the number of received states to be decoded, there will be a trade-off between the reduction in noise margin and the improvement in received bit-error-rate. This is the so called 'coding gain' and can be as much as 100-fold improvement in error performance under ideal conditions.

Attention has been given to the use of one such redundant coding scheme - multilevel trellis coding [6] - in modems, which, under poor error-rate conditions, exhibits a positive improvement in overall performance. One version of this coding has been standardised for use in the V32 modem, where a convolutional encoding process generates a 5th bit to be added to the 4 information carrying bits, resulting in a 32 point constellation rather than the 16 point version in Fig. 2.9. The effect of this mapping produces a coding scheme which eliminates the occurrence of adjacent points in the constellation being used for successively transmitted transitions. In spite of the more complicated detection method often using the Viterbi maximum likelihood detection schemes [7], worthwhile benefits are claimed, particularly over digital ADPCM (adaptive differential pulse code modulation) channels, where quantisation and non-linear distortions present severe problems for the voiceband modem. Over more conventional linearly distorted telephone channels, the benefits are less pronounced.

The V32bis recommendation for a 14400bit/s modem has now been issued, and is a logical extension to V32. The constellation definitions for the new speeds of 1200bit/s and 14400bit/s are taken from V33 and use trellis coding for all the higher speeds. The lowest speed of 4800bit/s is retained as the training speed for simplicity and backwards compatibility with V32. In addition to the speed range the new recommendation includes a method of dynamically changing speed depending on the prevailing line conditions. This reduces the error rate under adverse conditions by down-speeding and subsequently up-speeding when the conditions improve. The duration of the speed change has been kept short by using a 'rate renegotiation' algorithm which changes the speed at a rate which is within the error correction retransmit time scale. This makes the speed change transparent to the user who cannot distinguish between retrains and renegotiations. Rate adaptation between the modem and the user's DTE ensures that the interface is always at the same speed and same parity irrespective of the actual line rate.

Currently the CCITT working party are investigating extending the possible line rates to 19200bit/s, 24000bit/s and 28800bit/s in the so-called V.fast recommendation. This is due to be issued in the immediate future and will be allocated a number in place of 'fast' when it is released. This new modem will make even better use of the bandwidth available since the network is now mostly digital. However, it is expected to be backwards compatible with V32 and the other existing V series recommendations, modulations and protocols, and offer asynchronous rates that could be in direct competition with the 64kbit/s offered by ISDN.

2.6   The choice of modulation scheme

It would be tempting to tabulate the individual modulation schemes against various
network parameters and then select the best scheme for a given application. Such an
approach is , however, less than practical, as consideration of the choice of modulation
plan for a given bit rate is more likely to be influenced by ease of implementation and
by its upward or downward speed compatibility with other modems, as in the cases of
V22 and V22bis common at 1200bit/s, and the common speeds of 4800 and 9600bit/s
in V32 and V32bis. In a 'green field' situation, the classical method of comparing bit
error-rate performance with additive gaussian noise for different modulation and
detection methods is a tempting vehicle for analysis. However, although useful as a
yardstick for optimising the overall design, additive noise is one of the least onerous
impairments experienced on telephone circuits. Of equal importance is the ability of
both the modulation scheme and the modem itself to recover from the transient effects
of impulsive noise, phase hits, phase jitter, amplitude hits, line drop outs and the
transient effects of trunk switching.

Within these confines the CCITT sets out to agree an acceptable compromise.
Note that the CCITT Recommendations are literally recommendations in themselves
and are not mandatory requirements and are only intended to ensure compatibility
across different manufacturers' products. Individual PTT authorities are of course at
liberty to impose extra regulations and this can be seen in Europe with the attempt to
harmonise the requirements to a common specification (NET 20). Once a
recommendation is established, the performance of a particular manufacturer's modem
product is generally more influenced by his ability to provide jitter tracking, stable
automatic gain control circuits, optimised attenuation equaliser and echo cancellation
algorithms etc., than in his scope for manipulating the modulation process.

2.7   Equalisation

But what of the effect of group delay and amplitude distortion mentioned earlier? It has
already been noted that this will produce intersymbol interference in the received data
signal. To a first approximation, this could be eliminated by placing an artificial
network or equaliser, having the inverse characteristics of the telephone channel, in the
modem receiver. In practice, however, each telephone connection will have a different
set of channel characteristics, requiring an infinite number of equaliser characteristics
to be compensated. On some modems, attempts to approximate this approach have
been made by providing one or more manually selectable equalisers to effect some
form of Compromise Equaliser. This arrangement was favoured for use on the private
circuit connection (leased line), where the overall network characteristics are better
controlled and where the connection, once established, is seldom changed.

Where this approach is not acceptable, i.e. on most high speed modems on the
dial-up PSTN network, the solution is provided by the use of an Automatic or
Adaptive Equaliser, which automatically adjusts itself to the network passband
characteristics, and, in addition, compensates for the effects of echoes (delayed
facsimiles of the signal arising from reflections within the network).

A detailed exposition of adaptive equalisation is beyond the scope and space
limitations of this chapter (see [8] et al.) but, by way of an outline description, it might
take the form of the transversal filter shown in Fig. 2.10. Samples of the received line
signal are passed through a delay line having a spacing at the symbol or a sub-multiple
of the symbol period. By a suitable algorithm, the derived error signal can be used to
add or subtract from the pre- and post-distortion waveforms of successive line signals
(pulses) by automatically adjusting their contribution, through control of the variable

gain taps. Initial adjustment of the equaliser is achieved by transmitting a prescribed special conditioning or 'training' sequence, known to the receiving modem, at the start of each connection. The equaliser in the receiver, during this sequence, adjusts itself to minimise the effects of line distortion in the time domain; thereafter, the equaliser continually updates (or adapts) itself to the received data signal, to provide 'fine tuning'.

In principle this would also compensate for any time variance in the channel, although variation in time is not usually experienced, given the relatively short duration of most data calls. The 'training' sequences are defined by the relevant CCITT Recommendation (V29,V22,V32 etc.) but the design parameters of the equaliser are left to the discretion of the manufacturer, who will trade performance against economy of design.

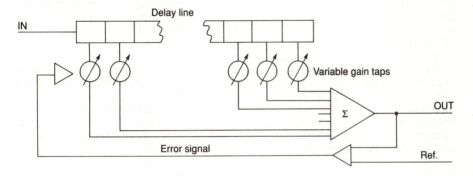

Fig. 2.10   *The transversal filter as an equaliser*

2.8   Scramblers

Illustrations of data signal spectra, such as Fig. 2.5, generally assume that the generating data sequence is of a random nature; were it not, we would see a less uniform spectrum with pronounced peaks. Not only can such peaks cause interference to adjacent channels and services, the lack of randomness in the data sequence hinders correct operation of adaptive equalisation, timing and carrier recovery circuits, especially those used in higher speed synchronous modems.

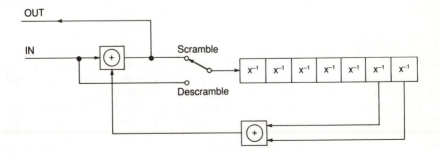

Fig. 2.11 *A typical scrambler arrangement*

In practice, data is seldom truly random and often contains strings of repeated patterns, which the modem randomises by scrambling the incoming bit stream, using a shift register arrangement implemented in either software or hardware and similar to that shown in Fig. 2.11. A complementary self-synchronising descrambler is employed in the receiver. For compatibility reasons, the scrambling algorithm polynomials are specified in the appropriate CCITT Recommendation.

2.9   Two-wire duplex modems

Duplex operation, the simultaneous transmission of data in both directions over 2-wire telephone circuits, necessitates the separation of the two directions of transmission. For voiceband modems this is usually implemented in one of two accepted ways.

2.9.1   Frequency Division Multiplexing

The most obvious approach uses Frequency Division Multiplexing (FDM), in which the telephone bandwidth is divided into two separate go and return frequency bands or channels. FDM is used for V21 (300bit/s), V22 (1200bit/s) and V23 (1200/75bit/s) to good effect but, for higher speeds, the problem of the inefficient use of the bandwidth becomes acute; for example the V22bis (2400bit/s) modem uses this method and requires a 16 point constellation for each channel, whereas a full bandwidth signal could achieve 9600bit/s with the same constellation, as shown in Fig. 2.9.

2.9.2   Echo cancellation

For higher speeds, the full use of the bandwidth is a prerequisite for detecting the required multilevel constellations and the designing of adequate channel separation filters becomes the limiting factor. FDM is therefore usually replaced by Echo Cancelling (EC). With this method, the full telephone bandwidth is employed in both directions of transmission and adaptive cancelling circuitry is used in the receiver to cancel the interference from both its own transmitter and the returned echoes from the network. A system of automatic and adaptive echo cancelling is used, much in the same way as for the adaptive equalisers described above.

Echo cancellation in modems has received most recent exposure through the 9600bit/s V32 recommendation, but an earlier V26ter echo canceller for 2400bit/s DPSK modulation was implemented by some manufacturers, although it has received less support than the alternative V22bis FDM implementation (due mainly to the latter's commonality with V22).

2.10   Filtering

If one were to look at some of the earlier designs, much of the hardware would be seen to consist of analogue filters. These would be mainly line and baseband filters, associated with the modem transmit and receive functions. The modem would also include other filters in the timing and carrier recovery circuits. Pulse shaping and a raised cosine response would be included in the filtering and has been shown to be a necessary requirement for the overall channel shaping. By applying the concept of matched filtering, [10] et al., necessary filtering should be split equally between the modem transmitter and receiver circuitry. The theory of detecting signals in the presence of gaussian noise is beyond the scope of this chapter but see references [9] and [10].

In addition to the pulse shaping, the transmitter filter will also be needed to attenuate any unwanted high-order modulation products, whilst the receiver filters will provide immunity to other unwanted out-of-band signals which might appear on the network. Often, the opportunity is taken of including a degree of pre-emphasis or de-emphasis shaping of the transmitted and received spectrum, as a form of compensation for channel distortion as indicated in 2.7 above. In the earlier modems, these filters would appear as complex active or passive implementations but today most of the shaping and filtering uses digital filtering as part of the digital signal processing function for both modulation and detection and only a small analogue filter is required external to the signal processor.

2.11 Data coding and error correction

With any modem, there is a high probability of errors occurring in the received data because of both long term noise levels on the telephone networks and from the shorter periods of noise bursts. To reduce the effects of this, error detection and protection methods are introduced into the data stream itself. In its simplest form this might involve packaging the data sequence into short blocks, to each of which are added additional check or parity bits. The remote end then uses these additional bits to confirm the integrity of each received block, any discrepancy resulting in a request for retransmission of the block in error.

The purpose of any error detection method is to eliminate the effect of line errors. However, the additional bits must be a balance between loss of absolute maximum data throughput in good conditions and gain in accurate data under adverse conditions. Small size packets are less likely to be corrupted but have a high overhead percentage, whereas large packets have less redundancy but are only viable when the basic transmission link error-rate is very low. Some protocols have adaptive packet sizes to cover the normally expected link error-rate and are usually in the range 64bit to 1024bit and can adapt in size depending on the prevailing link error-rate.

In recent years, attempts to improve the transmission efficiency have resulted in proposals for more complex coding schemes which combine error correction with data compression. CCITT Recommendation V42 covers two types of error correction, firstly the 'LAP M' protocol which is an extension of the HDLC/X25 Level 2 error correction mechanism, and secondly the Microcom Networking Protocol (MNP), [2] which has been widely used in the existing installed base of V22bis modems. Data compression needs an accurate error-free path to propagate its protocol and each of the V42 alternatives supports a specific algorithm; V42 'LAP M' supports V42bis and V42 'MNP 4' supports MNP 5.

Data compression is obtained by ensuring that the most frequently transmitted characters are sent by a shortened code. For example, an average textual message, such as a letter or memorandum contains many blank spaces and lines, often referred to as 'white space'. These can be sent more efficiently by transmitting a short code to indicate a repeated character or other often occurring words. This is the method used by MNP 5 and gives a compression of about 2:1 under favourable textual input. V42bis goes a stage further and uses a dynamic self-generating dictionary of strings of characters and then subsequently uses a short code if the string occurs again. The dictionary is normally of about 2,000 elements and is continually updated as the data transfer continues. Compression of about 4:1 can be achieved with this algorithm, assuming a suitable textual input. Conversely, a file with no redundancy and random non-repetitive string would be transmitted almost unchanged (e.g. an encrypted file). Under these circumstances the compression may be disabled deliberately by command code sequences within the data stream.

2.12   Modem control functions

The standard interface between the modem and the user's terminal, computer etc., collectively referred to as the Data Terminating Equipment (DTE), is governed by CCITT Recommendation V24, which defines a series of interchange circuits covering the exchange of data, timing signals and control information. Although V24 covers a wide range of circuits to cater for various applications, the most commonly used are shown on Table 2.2.

| V24 Circuit | Description | Source | |
		Modem	DTE
102	Signal ground/Common return	X	X
103	Transmitted data (TD)		X
104	Received data (RD)	X	
105	Request to send (RTS)		X
106	Clear to send (CTS)	X	
107	Data set ready (DSR)	X	
108/1or2	Connect to line or Data term.ready (DTR)		X
109	Received line signal (DCD)	X	
113	Tx signal element timing (DTE source)		X
114	Tx signal element timing (modem source)	X	
115	Rx signal element timing	X	
125	Calling indicator	X	

Table 2.2 *Some commonly used CCITT V24 interchange circuits*

More recently, the growth in integral card modems for terminal and personal computer (PC) equipments has seen the introduction of a new command structure or language, which enables the modem and DTE to communicate directly across the data transmit/receive interface circuits, for the purposes of configuration and control. Similar in concept to V25bis (see 2.12.2), the new procedures extend beyond the automatic call/answer commands to cover other features such as diagnostics, call status, speed changes, operating modes etc. As yet there is no international CCITT standard for this approach, although the market is tending towards the de-facto 'AT' command set originated by Hayes [11] in the USA.

2.12.1   Automatic modem connection to line

Originally, modems were manually connected to PSTN lines in both originate (calling) and answer (called) modes, first by establishing a speech path, using a telephone instrument, and then manually switching the modem to line. An early development was the provision of an Automatic Answer (AA) facility at the called modem whereby the modem, on receipt of the incoming ringing, automatically connected to line, permitting ready access to unattended bureaux and databases. Later developments allowed for the complementary feature of originating calls automatically, under DTE control, from an external unit, which received its calling instructions across a separate parallel interface, known as the V24 200-series interchange circuits. These external units were placed between the DTE and modem and carried out all of the call set-up routines, including dialling out to line. Once the connection was established the autocaller connected the modem to the line and handed control over to the normal modem/DTE interface. The equipment involved was bulky and relatively expensive and was only used in specialist applications.

With the arrival of microprocessor-based control circuitry, the potential for providing greater intelligence within the modem lent itself to the provision of both automatic answering and automatic calling within the modem for very little additional cost. Unfortunately, the drawback of still having to provide the 200-series interface remained. Spurred on by the advantage of auto-calling in many data communications applications, there emerged various schemes for processing the serial control and number entry information across the existing transmit and receive interchange circuits (circuits 103 and 104).

2.12.2 Modem control using the serial data interface

As these schemes proliferated, it became clear that there was a need for an international standard, from which arose the CCITT Recommendation V25bis for 'Automatic Calling and/or Answering Equipment on the GSTN using the 100-series Interchange Circuits', as first published in the 'Red' book during 1985. In addition to defining the command and response indications, V25bis also covers telephone number storage, dialling, repeat call attempts and maintenance aspects. Thus, by differentiating between the periods of when the data terminal 'talks' to the modem for control purposes and those when the modem is in its normal mode of transferring data to line, the terminal can control what function the modem should perform, prior to and even during normal modem operation.

This method is now preferred to all of the previous methods and V25bis is quite common throughout Europe. However, in the USA, the Hayes 'AT' protocol [11] has become a de-facto standard and, although imperfectly defined, its concept allows extension and expansion to cover command structures other than dialling. Commonly it now includes configuration of the modem as far as the 'V' standard, error correction, data compression and other interactive actions between the modem and the DTE. This has become extensively used with personal computers (PC) terminals with software packages designed to interface with the modem using the 'AT' command set philosophy. CCITT are considering adopting the same basic principles in the interface between the DTE and the modem in a new recommendation extending the command structures beyond V25bis.

2.13 Diagnostic features

Although most large data networks now provide an independent Network Management and/or Control capability, there still remains a need for a simple diagnostic facility within the modem itself, enabling the user to quickly determine whether the source of any problem is attributed to one or other of the network, modem or terminal equipment. The basic requirements are formulated in CCITT Recommendation V54 and comprise, essentially, the two main test loops, as shown in Fig. 2.12.

In addition to, and in conjunction with, these basic requirements, extra features such as the provision of internal pattern generation and checking, addressing and status monitoring are often included by the manufacturer. These facilities lend themselves to control and interrogation by the terminal, using the serial interface, as described in 2.12.2 above.

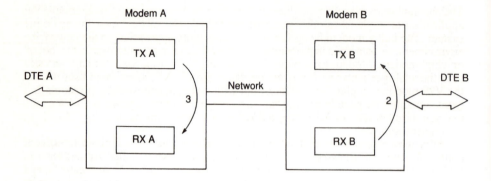

Fig. 2.12 *Diagnostic Features*

LOOP 2   Digital Loop - loops the data received by Modem B and
         retransmits it for receipt and checking by Modem A

LOOP 3   Analogue Loop - loops the analogue transmit and receive
         signals at the local modem

2.14   Line interfacing

It would be inappropriate to conclude a paper on modems without a passing reference
to the line interface circuitry, which provides not only a convenient means of matching
the modem to the telephone network but also affords the essential electrical safety
barrier between the network and the 'modem plus terminal' (and vice versa) required
by PTTs and national administrations. In the UK, control of these requirements has
passed from British Telecom to the Department of Trade and Industry (Office of
Telecommunication - OFTEL). This control is in the form of British Standards (BS)
specifications which outline the parameters associated with both user and network
protection. At present, all modems, as with other attachments to the network, must be
submitted for testing through the British Approvals Board for Telecommunications
(BABT) before permission can be given for connection to networks licensed in
accordance with the prevailing Telecommunication Act.

It is perhaps worth noting that, within Europe, a process is in place to establish a
series of standards or NETs (Normes Européens de Télécommunications), covering
both safety and performance requirements. These standards are being formulated with
the objective of enabling a common set of product approval criteria, which would be
recognised across all member states. An attempt to harmonise the analogue parameters
across member states is also under consideration in NET 4, despite the differences in
implementation of each different country's networks, where some common parameters
have been identified.

2.15 The future

The past few years have seen fundamental changes in the evolution of modems and the application of modems to other products for the transfer of data over the analogue system. The bulky modem has given way to units more appropriate to the new office environment, in both style and operation, with an increasing trend towards the 'board' or 'chip' modem. Modem technology is continuing to be used for other data services over the analogue network. For example, facsimile (FAX) transmission uses mainly the V29 modulation plan for 9600bit/s, but with different handshaking protocols and extra signalling for page boundaries and headers etc. A recent update to a 14,400bit/s fax standard, based on V32bis modulation, has been introduced by CCITT as recommendation V17.

Microprocessors have enabled the modem manufacturer to include intelligent control features as discussed in 2.12, at little extra cost, whilst other features such as built-in error correction (fast becoming a mandatory selling point), automatic speed and V series recognition and user configurable software are now appearing as standard. Surprisingly, through the use of new components and manufacturing techniques, all this has been achieved at the same time as the cost has been reduced to nearly 30% of its price ten years ago. Fifteen or twenty years ago, even the simplest modem was realised from discrete analogue circuitry. By the late 1970s, much of the analogue circuitry had given way to an increasing amount of digital circuitry and, by the early 1980s, the microprocessor and purpose-built digital signal processors (DSPs) had covered virtually all of the modulation, demodulation, filtering and equalisation functions.

More recently, the original business market for modems has been enlarged by the growth in private users. This follows the trend in the entry of the personal computer (PC) into the home environment; in fact there are now on offer integral and stand-alone modems to cover both business and domestic applications using the same hardware. For speeds up to 2400bit/s full duplex (V22bis) the 'modem on a chip' is already established, for just a few pounds sterling, as the heart of most low speed applications. For the higher speed offerings the DSP has become dominant and this has led to the semiconductor manufacturers becoming an important influence on the evolution of new modem technology. Another innovation is the transfer of video information across the analogue network and this has implications for both the business user for multi-media communication PC to PC and in the domestic environment for a videophone as recently introduced by BT. This is based on V32bis modem technology and uses the 14,400bit/s digital bandwidth divided as 60% for video, 30% for speech and 10% for control.

How much all these trends will continue is difficult to predict, and much will depend on the speed of transition to the all-digital ISDN network. Already semiconductor manufacturers are producing sets of integrated circuits, which can provide all of the necessary functions of echo cancelling, digital transmission, multiplexing, signalling and interfacing to the ISDN circuit. Once the problems of physical packaging and cost have been resolved, the 'digital' modem could provide the single access module between the customer's terminal and the digital network. Balancing this trend will be the next phase in analogue technology with the introduction of the new 'V.fast' recommendation from CCITT, which will take the line rate to 28,800bit/s, and with V42bis compression will give an asynchronous transfer rate of 115,200bit/s, a definite competitor for the basic ISDN rate of 64,000bit/s.

Notwithstanding all of this, whilst the modem-based products will continue to evolve, it would be premature to dismiss the traditional analogue modem this side of the 21st century.

2.16   References

1.   International Telecommunications Union - Telegraph and Telephone Committee (CCITT) 'Data Communication Over the Telephone Network', Recommendations of the V Series. CCITT Blue Book VIII.I

2.   MNP - Microcom Networking Protocol - A proprietary protocol developed by Microcom Inc, Boston USA

3.   Duc, N.Q. and Smith, B.M., 'Line Coding for Digital Data Transmission', *Australian Telecom Review* , 1977, **II**, pp.14-27

4.   Shannon, G.E., 'A Mathematical Theory of Communication', *BSTJ* , 1948, **27**

5.   Nyquist, H., 'Certain Topics in Telegraph Transmission Theory', *Transactions AIEE,* 1928, **47**

6.   Ungerbeck, G., 'Channel Coding with Multilevel Phase Signals', *IEEE Transactions Info.Theory,* 1982, **IT28** No1, and Wei, L.F., 'Rotationally Invariant Convolutional Channel Coding with Expanded Signal Space', *IEEE Transactions Selected Areas in Communications*, 1984, **SAC-2** No5, p. 661

7.   Forney, G.D., 'The Viterbi Algorithm', *Proc IEEE*, 1973, **61**

8.   Lucky, R.W., Salz, J. and Weldon, E.J., 1968, 'Principles of Data Communication' (McGraw-Hill)

9.   Clark, A.P., 1976, 'Principles of Digital Data Transmission' (Pentech Press)

10. Bennett, W.R. and Davey, J.R., 1965, 'Data Transmission' (McGraw-Hill)

11. Hayes AT command set - A proprietary command structure developed by Hayes Microcomputer Products Inc., Georgia, USA

Chapter 3

The philosophy of the OSI seven-layer model

Fred Halsall

3.1   Computer communications requirements

Although in many instances computers are used to perform their intended role in a stand-alone mode, in others there is a need to interwork and exchange data with other computers. In financial applications, for example, to carry out funds transfers from one institution computer to another, in travel applications to access the reservation systems belonging to various airlines, and so on. The general requirement in all these applications is for application programs running in different computers to cooperate to achieve a specific distributed application function. To achieve this, three basic issues must be considered. These are shown in diagrammatic form in Fig. 3.1.

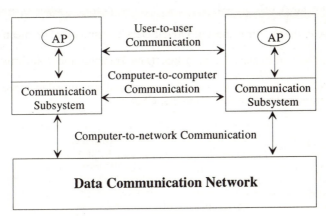

AP = Application Process

Fig. 3.1 *Computer communication schematic*

The fundamental requirement in all applications that involve two or more computers is the provision of a suitable data communications facility. This may comprise a local area network if the computers are distributed around a single site, a wide area network if the computers are situated at different sites or an internetwork if multiple interconnected network types are involved.

Associated with these different network types is a set of access protocols which enable a communications path between two computers to be established and for data to be transferred across this path. Typically these protocols differ for the different network types. In addition to these access protocols, the communication subsystem in each computer must provide additional functionality. For example, if the communicating computers are of different types, possibly with different word sizes and character sets, then a means of ensuring the transferred data is interpreted in the same way in each computer must be incorporated. Also, the computers may use different file systems and hence functionality to enable application programs - normally referred to as application processes or APs - to access these in a standardised way must also be included. All these issues must be considered when communicating data between two computers.

3.2   Standards evolution

Until recently, the standards established for use in the computer industry by the various international bodies were concerned primarily with either the internal operation of a computer or the connection of a local peripheral device. The result was that early hardware and software communication subsystems offered by manufacturers only enabled their own computers, and so-called plug-compatible systems, to exchange information. Such systems are known as closed systems since computers from other manufacturers cannot exchange information unless they adhere to the (proprietary) standards of a particular manufacturer.

In contrast, the various international bodies concerned with public-carrier telecommunication networks have for many years formulated internationally agreed standards for connecting devices to these networks. The V-series recommendations, for example, are concerned with the connection of equipment - normally referred to as a data terminal equipment (DTE) - to a modem connected to the PSTN; the X-series recommendations for connecting a DTE to a public data network; and the I-series recommendations for connecting a DTE to the emerging ISDNs. The recommendations have resulted in compatibility between the equipment from different vendors, enabling a purchaser to select suitable equipment from a range of manufacturers.

Initially, the services provided by most public carriers were concerned primarily with data transmission, and hence the associated standards only related to the method of interfacing a device to these networks. More recently, however, the public carriers have started to provide more extensive distributed information services, such as the exchange of electronic messages (Teletex) and access to public databases (Videotex). To cater for such services, the standards bodies associated with the telecommunications industry have formulated standards not only for interfacing to such networks but also so-called higher level standards concerned with the format (syntax) and control of the exchange of information (data) between systems. Consequently, the equipment from one manufacturer that adheres to these standards can be interchangeable with equipment from any other manufacturer that complies with the standards. The resulting system is then known as an open system or, more completely, as an open systems interconnection environment (OSIE).

In the mid 1970s as different types of distributed systems (based on both public and private data networks) started to proliferate, the potential advantages of open systems were acknowledged by the computer industry. As a result, a range of standards started to be introduced. The first was concerned with the overall structure of the complete communication subsystem within each computer. This was produced by the International Standards Organization (ISO) and is known as the ISO Reference Model for Open Systems Interconnection (OSI).

The aim of the ISO Reference Model is to provide a framework for the coordination of standards development and to allow existing and evolving standards activities to be set within a common framework. The aim is to allow an application process in any computer that supports a particular set of standards to communicate freely with an application process in any other computer that supports the same standards, irrespective of its origin of manufacture.

Some examples of application processes that may wish to communicate in an open way are:

◊   a process (program) executing in a computer and accessing a remote file system;
◊   a process acting as a central file service (server) to a distributed community of (client) processes;
◊   a process in an office workstation (computer) accessing an electronic mail service;
◊   a process acting as an electronic mail server to a distributed community of (client) processes;
◊   a process in a supervisory computer controlling a distributed community of computer-based instruments or robot controllers associated with a process or automated manufacturing plant;
◊   a process in an instrument or robot controller receiving commands and returning results to a supervisory system;
◊   a process in a bank computer that initiates debit and credit operations on a remote system.

Open systems interconnection is concerned with the exchange of information between such processes. The aim is to enable application processes to cooperate in carrying out a particular (distributed) information processing task irrespective of the computers on which they are running.

3.3   ISO Reference Model

A communication subsystem is a complex piece of hardware and software. Early attempts at implementing the software for such subsystems were often based on a single, complex, unstructured program (normally written in assembly language) with many interacting components. The resulting software was difficult to test and often very difficult to modify.

To overcome this problem, the ISO has adopted a layered approach for the reference model. The complete communication subsystem is broken down into a number of layers each of which performs a well defined function. Conceptually, these layers can be considered as performing one of two generic functions; network-dependent functions and application-oriented functions. This in turn gives rise to three distinct operational environments:

1. The network environment, which is concerned with the protocols and standards relating to the different types of underlying data communication networks.

2. The OSI environment, which embraces the network environment and adds additional application-oriented protocols and standards to allow end systems (computers) to communicate with one another in an open way.

3. The real systems environment, which builds on the OSI environment and is concerned with a manufacturer's own proprietary software and services which have been developed to perform a particular distributed information processing task.

This is shown in diagrammatic form in Fig. 3.2.

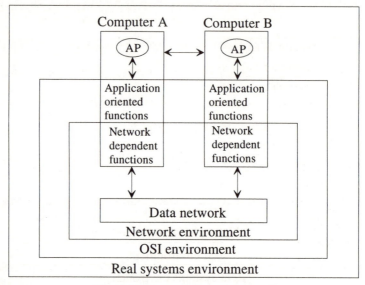

Fig. 3.2 *Operational environments*

Both the network-dependent and application-oriented (network-independent) components of the OSI model are implemented as a number of layers. The boundaries between each layer, and the functions performed by each layer, have been selected on the basis of experience gained during earlier standardisation activity.

Each layer performs a well defined function in the context of the overall communication subsystem. It operates according to a defined protocol by exchanging messages, both user data and additional control information, with a corresponding peer layer in a remote system. Each layer has a well defined interface between itself and the layer immediately above and below. Consequently, the implementation of a particular protocol layer is independent of all other layers.

The logical structure of the ISO Reference Model is made up of seven protocol layers as shown in Fig. 3.3.

The three lowest layers (1-3) are network-dependent and are concerned with the protocols associated with the data communication network being used to link the two communicating computers. In contrast, the three upper layers (5-7) are application-oriented and are concerned with the protocols that allow two end user application processes to interact with each other, normally through a range of services offered by the local operating system. The intermediate transport layer (4) masks the upper application-oriented layers from the detailed operation of the lower network-dependent layers. Essentially, it builds on the services provided by the latter to provide the application-oriented layers with a network-independent message interchange service.

The function of each layer is specified formally as a protocol that defines the set of rules and conventions used by the layer to communicate with a similar peer layer in another (remote) system. Each layer provides a defined set of services to the layer immediately above. It also uses the services provided by the layer immediately below

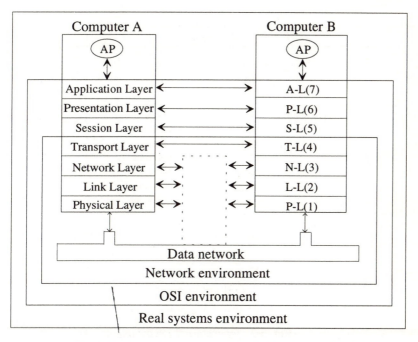

Fig 3.3   *Overall structure of the ISO Reference Model*

it to transport the message units associated with the protocol to the remote peer layer.
For example, the transport layer provides a network-independent message transport
service to the session layer above it and uses the service provided by the network layer
below it to transfer the set of message units associated with the transport protocol to a
peer transport layer in another system.   Conceptually, therefore, each layer
communicates with a similar peer layer in a remote system according to a defined
protocol.   However, in practice the resulting protocol message units of the layer are
passed by means of the services provided by the next lower layer.   The basic functions
of each layer are summarised in Fig. 3.4.

3.3.1 The application-oriented layers

3.3.1.1 The application layer
The application layer provides the user interface - normally an application
program/process - to a range of network-wide distributed information services.   These
include file transfer access and management as well as general document and message
interchange services such as electronic mail.   A number of standard protocols are either
available or are being developed for these and other types of service.
    Access to application services is normally achieved through a defined set of
primitives, each with associated parameters, which are supported by the local
operating system.   The access primitives are the same as other operating system calls
(as used for access to, say, a local file system) and result in an appropriate operating
system procedure (process) being activated. These operating system procedures use

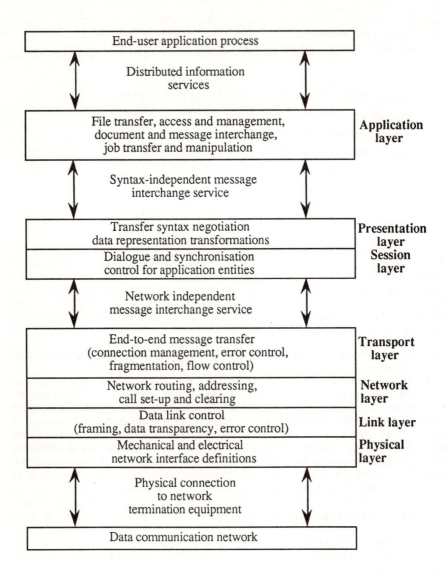

Fig 3.4   *Protocol layer summary*

the communication subsystem (software and hardware) as if it is a local device - similar to a disk controller, for example. The detailed operation and implementation of the communication subsystem is thus transparent to the (user) application process. When the application process making the call is rescheduled (run), one (or more) status parameters are returned indicating the success (or otherwise) of the network transaction that has been attempted.

In addition to information transfer, the application layer provides such services as:

◊ identification of the intended communication partner(s) by name or by address;

◊ determination of the current availability of an intended communication partner;

◊ establishment of authority to communicate;

◊ agreement on privacy (encryption) mechanisms;

◊ authentication of an intended communication partner;

◊ selection of the dialogue discipline, including the initiation and release procedures;

◊ agreement on responsibility for error recovery;

◊ identification of constraints on data syntax (character sets, data structures, etc.)

3.3.1.2 The presentation layer

The presentation layer is concerned with the representation (syntax) of data during transfer between two communicating application processes. To achieve true open systems interconnection, a number of common abstract data syntax forms have been defined for use by application processes together with associated transfer (or concrete) syntaxes. The presentation layer negotiates and selects the appropriate transfer syntax(es) to be used during a transaction so that the syntax (structure) of the messages being exchanged between two application entities is maintained. Then, if this form of representation is different from the internal abstract form, the presentation entity performs the necessary conversion.

To illustrate the services provided by the presentation layer, consider a telephone conversation between a French speaking person and a Spanish speaking person. Assume each uses an interpreter and that the only language understood by both interpreters is English. Each interpreter must translate from their local language to English, and vice versa. The two correspondents are thus analogous to two application processes with the two interpreters representing presentation layer entities. French and Spanish are the local syntaxes and English the transfer or concrete syntax. Note that there must be a universally understood language which must be defined to allow the agreed transfer language (syntax) to be negotiated. Also note that the interpreters do not necessarily understand the meaning (semantics) of the conversation.

Another function of the presentation layer is concerned with data security. In some applications, data sent by an application is first encrypted (enciphered) using a key, which is (hopefully) known only by the intended recipient presentation layer. The latter decrypts (deciphers) any received data using the corresponding key before passing it on to the intended recipient.

3.3.1.3 The session layer

The session layer provides the means that enables two application layer protocol entities to organise and synchronise their dialogue and manage their data exchange. It is thus responsible for setting up (and clearing) a communication (dialogue) channel between two communicating application layer protocol entities (presentation layer protocol entities in practice) for the duration of the complete network transaction. A number of optional services are provided, including:

◊   Interaction management: The data exchange associated with a dialogue may be duplex or half-duplex. In the latter case it provides facilities for controlling the exchange of data (dialogue units) in a synchronised way.

◊   Synchronisation: For lengthy network transactions, the user (through the services provided by the session layer) may choose periodically to establish synchronisation points associated with the transfer. Then, should a fault develop during a transaction, the dialogue may be restarted at an agreed (earlier) synchronisation point.

◊   Exception reporting: Non-recoverable exceptions arising during a transaction can be signalled to the application layer by the session layer.

3.3.1.4 The transport layer

The transport layer acts as the interface between the higher application-oriented layers and the underlying network-dependent protocol layers. It provides the session layer with a message transfer facility that is independent of the underlying network type. By providing the session layer with a defined set of message transfer facilities the transport layer hides the detailed operation of the underlying network from the session layer.

The transport layer offers a number of classes of service which cater for the varying quality of service (QOS) provided by different types of network. There are five classes of service ranging from:

◊   class 0, which provides only the basic functions needed for connection establishment and data transfer, to

◊   class 4, which provides full error control and flow control procedures.

As an example, class 0 may be selected for use with a packet-switched data network (PSDN) while class 4 may be used with a local area network (LAN) providing a best-try service; that is, if errors are detected in a frame then the frame is simply discarded.

3.3.2   The network-dependent layers

As the lowest three layers of the ISO Reference Model are network dependent, their detailed operation varies from one network type to another. In general, however, the network layer is responsible for establishing and clearing a network-wide connection between two transport layer protocol entities. It includes such facilities as network routing (addressing) and, in some instances, flow control across the computer-to-network interface. In the case of internetworking it provides various harmonising functions between the interconnected networks.

The link layer builds on the physical connection provided by the particular network to provide the network layer with a reliable information transfer facility. It is thus responsible for such functions as error detection and, in the event of transmission errors, the retransmission of messages. Normally, two types of service are provided:

1.   Connectionless, which treats each information frame as a self-contained entity that is transferred using a best-try approach.

2.   Connection-oriented, which endeavours to provide an error-free information transfer facility.

Finally, the physical layer is concerned with the physical and electrical interfaces between the user equipment and the network terminating equipment. It provides the link layer with a means of transmitting a serial bit stream between the two equipments.

3.4 Open system standards

The ISO Reference Model has been formulated simply as a template for the structure of a communication subsystem on which standards activities associated with each layer may be based. It is not intended that there should be a single standard protocol associated with each layer. Rather, a set of standards is associated with each layer, each offering different levels of functionality. Then, for a specific open systems interconnection environment, such as that linking numerous computer-based systems in a fully-automated manufacturing plant, a selected set of standards is defined for use by all systems in that environment.

The three major international bodies actively producing standards for computer communications are the ISO, the (American) Institute of Electrical and Electronic Engineers (IEEE) and the International Telegraph and Telephone Consultative Committee (CCITT). Essentially, ISO and IEEE produce standards for use by computer manufacturers while CCITT defines standards for connecting equipment to the different types of national and international public networks. As the degree of overlap between the computer and telecommunications industries increases, however, there is an increasing level of cooperation and commonality between the standards produced by these organisations.

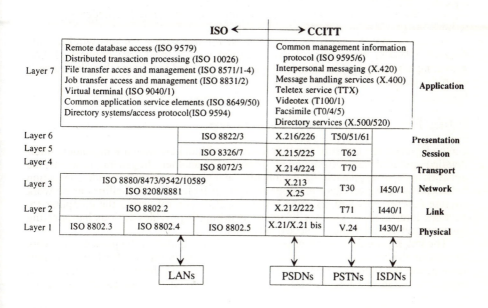

Fig. 3.5 *Standards summary*

A range of standards has been defined by the ISO/CCITT and a selection of these is shown in Fig. 3.5. Collectively they enable the administrative authority that is establishing the open system environment to select the most suitable set of standards for the application. The resulting protocol suite is known as the open system interconnection profile. A number of such profiles have now been defined, including: TOP, a protocol set for use in technical and office environments; MAP, for use in manufacturing automation; US and UK GOSIP for use in US and UK government projects, respectively; and a similar suite used in Europe known as the CEN functional standards. The latter has been defined by the standards Promotion and Application Group (SPAG), a group of twelve European companies.

As Fig. 3.5 shows, the lower three layers vary for different network types. CCITT has defined V-, X- and I-series standards for use with public-carrier networks. The V-series is for use with the existing switched telephone network (PSTN), the X-series for use with existing switched data networks (PSDN) and the I-series for use with the emerging integrated services digital networks (ISDN). Those produced by ISO/IEEE for use with local area networks are known collectively as the 802 (IEEE) or 8802 (ISO) series.

3.5   Summary

This chapter has reviewed the requirements for the communications subsystem in each of a set of interconnected computers that enables them to communicate in an open way to perform various distributed application functions. The philosophy behind the structure of the ISO Reference Model for open systems interconnection has been presented and a description of the functionality of the seven layers that make up the reference model described. Finally, a selection of the ISO/CCITT standards that have been defined have been identified.

3.6   Bibliography

Further coverage of the material covered in this chapter can be found in:

Halsall, F., 1992, "Data communications, computer networks and open systems", third edition (Addison Wesley Publishers)

Chapter 4

Data transmission standards and interfaces

John L. Moughton

4.1 Introduction - standards and interfaces

Standardisation is an essential aspect of data communications, as indeed it is of all communications. Communication between two people or pieces of equipment can succeed only if both parties obey an agreed set of rules. Whilst it is possible for a set of rules to be developed privately between the two parties and result in successful communication this is of no use if either party subsequently wishes to communicate with a third. It is clearly more useful if they follow a set of rules agreed by a larger population (a "standardised" set).

As an example, consider the case of a young baby communicating with his mother. Initially they will communicate by a set of verbal and non-verbal signals which will be meaningful only to themselves. As the child grows older he will start to learn a spoken language and discover that this allows him to communicate not only with his mother but also with many other people. He finds that there is a distinct advantage in using this "standardised" set of rules (language) as a means of communication.

Now imagine that our child is playing with a toy that can speak. In order for the words to be intelligible they must obviously be in a language known to the child and delivered at an appropriate rate and volume. The values of the three parameters - language, rate and volume - are therefore constrained, i.e. they need to conform to certain standards. There are, however, many ways in which a toy can be made to meet these requirements. For example, the toy could produce speech by purely mechanical means (acoustic gramophone), by reproducing magnetically recorded speech or by electronic synthesis. Indeed, different toys can use different techniques. Provided the standards are met at the boundary or interface (in this case the air between the toy and the child) then communication will be successful. Notice that the standards impose no *a priori* restrictions on the choice of technology. Any existing sound reproduction technique can be used and the choice can be made on performance and economic grounds. Also, the way is left open for any new, as yet undiscovered, method to be used.

This example shows that it is sufficent to standardise what happens at a boundary or interface. Indeed, it could be argued that since the air boundary is the only place where the performance of the toy can be measured, it would be meaningless to specify a standard elsewhere since that point could not be observed to verify conformance. Taking the toy apart to make a measurement would, in effect, be defining a new (test) interface.

4.2   What are standards?

A standard aims to achieve some form of compatibility or interoperability between items of equipment by specifying the technical characteristics necessary to achieve that end. An example of this is the standardised metric thread which is widely used for bolts and screws. The benefit of this standardisation has been to reduce the confusion that resulted from the many national and manufacturers' standards that existed in the past, for example, BSF, BSP, BA, UNF, UNC and Whitworth. Another example is the UK approval standard for telephone handsets which ensures that one can buy a telephone from any of a number of suppliers and be confident that it will work when plugged into a BT socket.

There are two different types of standards, categorised by how they are produced:

1. A *de jure* standard (literally *by right*) is set by a formally constituted standards organisation which has a certain geographical and/or organisational scope. It may be an international standard (covering the whole world) set by the ISO (International Organization for Standardization) or the ITU (International Telecommunication Union). It may be a regional (eg. European) or national (eg. British) standard, a standard set by a trade association, or even an internal company standard.

2. A *de facto* standard (literally *in fact*) is a standard which has been adopted informally by a community of interest. It is often a proprietary scheme which has been used because there is no *de jure* standard. An example of this is the well known AT command set for the control of dial-up asynchronous modems which was originally devised by a single modem manufacturer. Sometimes a *de facto* standard is adopted by a standards body and becomes *de jure*.

Recently a number of *industry fora* have sprung up. Each is concerned with promoting a particular technology. Examples are the Frame Relay Forum and the ATM Forum, both of which have large memberships mainly of manufacturers but also service providers and a few users. On the whole industry fora do not write new standards but rather adopt, possibly with modification, existing international or regional standards. However, in areas where there are no existing standards, a forum may start standards development, if only to provoke an official standards body into action.

The process of developing a standard is time consuming, particularly a *de jure* standard where many interests have to be considered. It may take five years or more and it is consequently an expensive process. *De facto* standards appear more quickly, having been devised by a single manufacturer and then being adopted more or less unchanged by others.

4.3   Why should we use standards?

There are those who question whether the benefits offered by standardisation actually outweigh the costs, however in the field of telecommunications there are a significant number of advantages:

1. Data communications standards permit systems (that is computers and terminals) from different suppliers to be interconnected and communicate successfully. These standards also permit equipment to be connected to public communications networks. An example of the importance of this is the facsimile (fax) machine. If no standards existed this communications medium would be little used since

only machines from the same manufacturer would be able to communicate. However, the existence of standards has made it one of the most important business communications tools in use today. There are many other examples, including the so-called V.24/RS232 interface, the V.21/V.23 modem and the X.25 interface specifcation.

2.   Standards can reduce a customer's dependence on a particular manufacturer since he can buy from any supplier whose products conform to the standard, with the confidence that they will interwork with his existing equipment. Some manufacturers deliberately do not use standards in order to lock a customer into a particular product. Alternatively they may provide conformance to basic standards (thus allowing simple interworking) whilst providing enhanced features (and thus a competitive edge) in a proprietary manner. Users of data communications equipment generally prefer to be free to buy from multiple suppliers since this both reduces cost (competition brings prices down) and reduces dependence on the fortunes of the supplier.

3.   Standards provide some guarantee of quality as the technical content has been reviewed by many experts from different organisations before being approved. Ideally, for data communications standards these will be equipment manufacturers, telecommunications service providers and users. Formal specification and verification techniques are being used in the development of standardised communications protocols which would be beyond the resources of most individual companies.

4.   International or regional standards can remove barriers to trade that would otherwise be set up by differing national practices. This is particularly significant in Europe as will be discussed later on.

5.   Safety and the protection of the environment may be ensured. Equipment for general use must meet stringent standards for electrical and mechanical safety and electromagnetic interference.

4.4 The organisation of the standards world

There are a number of different organisations operating at the international, regional and national levels. Over a period of time these have developed a way of working together which has resulted in a reasonably rationalised standards development process. This section introduces some of the key organisations in standards development.

4.4.1 International bodies

The International Organization for Standardization (ISO) was founded in 1947 and has its headquarters in Geneva. Its members are the national standardisation bodies of all the member countries. The British Standards Institution (BSI) represents the United Kingdom. The ISO is responsible for standardisation in all fields with certain exceptions where it has agreed that other organisations should be responsible (see IEC and ITU below).

The International Electrotechnical Commission (IEC) was founded in 1906 and has responsibility in the fields of electronic equipment, electrical engineering and industrial process control.

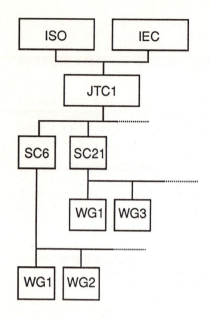

Sub-Committees

SC21 - Information retrieval, Transfer and Management for OSI.
Working Groups (WGs 1, 3, 4, 7, 8)

SC6 - Telecommunications and Information Exchange between Systems.
Working Groups (WGs 1-4, 6)

Fig. 4.1   *Structure of the IEC/ISO JTC1 Joint Technical Committee for Information Technology*

ISO/IEC Joint Technical Committee on Information Technology (JTC1). In 1987 the Information Technology (IT) standardisation activities of ISO and IEC were merged and put under the control of the IEC/ISO JTC1 joint technical committee. The structure (Fig. 4.1) consists of a number of sub-committees, each covering a different aspect of IT. Those most relevant to data communications are:

SC-6   Telecommunications and Information Exchange between Systems (excluding public telecommunications).

SC-6 has five Working Groups: WGs 1-4 cover Data Link, Network, Physical and Transport Layer standards respectively. WG6 covers Private ISDNs.

SC-21   Information Retrieval, Transfer and Management for Open Systems Interconnection (OSI).

SC-21 also has 5 WGs: 1 - OSI Architecture, 3 - Database, 4 - OSI Management, 7 - Open Distributed Processing (ODP) and 8 - OSI Upper Layers.

The International Telecommunication Union (ITU) was founded in 1865 as the International Telegraph Union and acquired its present name in 1932. It is a specialised agency of the United Nations which deals with public telecommunications. It is an organisation, a union, of Member countries with headquarters in Geneva. In October 1991 there were 164 members. The UK representative is the Department of Trade and Industry (DTI). The members send delegates to a Plenipotentiary Conference held every five years to revise the ITU Convention and elect key officers.

The functions of the ITU are as follows:

1.   To maintain and extend international cooperation for the improvement and rational use of telecommunications of all kinds.

2.   To promote the development of technical facilities and their most efficient operation with a view to improving the efficiency of telecommunications services, increasing their usefulness and making them as far as possible generally available to the public.

3.   To harmonise the actions of nations in the attainment of these common ends.

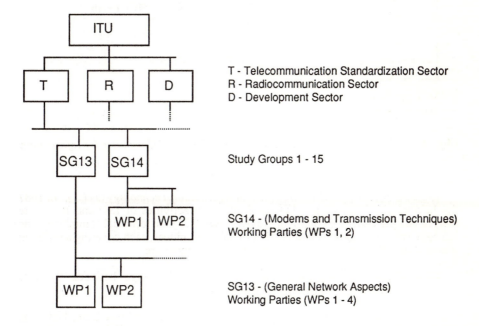

Fig. 4.2 *Structure of the ITU Telecommunication Standardization Sector (ITU-T)*

There have been a number of changes in the organisation of the ITU since its founding, the most recent being in December 1992. At this "Additional Plenipotentiary Conference" a new Constitution was agreed which, although formally taking effect from 1st July 1994, has, in fact, been applied from 1st March 1993. The new structure (Fig. 4.2) is organised into three sectors: development, standardisation and radiocommunication. The standards-writing activities of the CCITT (International Telegraph and Telephone Consultative Committee) and CCIR (International Radio Consultative Committee) have been merged into a Telecommunication Standards Sector (ITU-T). The remaining CCIR activities (mainly those concerning the efficient management of the radio-frequency spectrum) have been merged with those of the IFRB (International Frequency Registration Board) to form the Radiocommunication Sector. The BDT (Telecommunications Development Bureau) remains essentially the same but with increased emphasis on its role of helping developing countries develop policies and structures to improve their telecommunications development.

All ITU member countries can participate in the work of the ITU-T. A country's representatives can be drawn from its Telecommunications Administration, Recognised Private Operating Agencies and Scientific or Industrial Organisations. Also International Organisations are allowed to send delegates. The ITU-T will hold a World Telecommunication Standardization Conference (WTSC) every four years (replacing the CCITT's and CCIR's Plenary Assemblies). The WTSC draws up a list of technical telecommunications subjects or "Questions", the study of which would lead to improvements in international telecommunications. These Questions are entrusted to a number of Study Groups, composed of experts from the member organisations. The WTSC also approves the Recommendations (standards) produced during the previous four year period. There is in addition an accelerated approval procedure which can be used between WTSC meetings. Study Groups meet once or twice a year to progress the work of drafting new and revising old Recommendations and after each meeting a report is published. Amongst other things this contains updated texts of Recommendations and liaison statements (letters) to other Study Groups and standards bodies asking or answering questions and requesting or providing information.

The Study Groups with the most relevance to data communications are:

SG 7 Data networks and open systems communication (X-series Recommendations) including packet and circuit switched public data networks and Open Systems Interconnection (OSI).

SG 8 Terminals for telematic services (T-series Recommendations) including facsimile, videotex and document architectures.

SG 13 General network aspects (I- and some G-series Recommendations) including digital hierarchies and ISDN (64 kbit/s and Broadband) architectures, reference models and interfaces. (This SG was previously known as SG XVIII.)

SG 14 Modems and transmission techniques for data, telegraph and telematic services (V-series Recommendations) including modems, ISDN terminal adaptors and telegraph equipment. (This SG was formed by the merging of SGs IX and XVII.)

4.4.2 European standards organisations

The relationship between the European Economic Community (EEC) and telecommunications standardisation is discussed in a separate section. The key European standards organisations are described below.

The Comité Européen de Normalisation (CEN) is the main standards body in Europe and has the same scope as the ISO. It was founded in 1961 and comprises the national standards bodies of the EEC and EFTA (European Free Trade Association) countries.

The Comité Européen de Normalisation Électrotechnique (CENELEC) was founded in 1973 and is the European equivalent of the IEC, being responsible for electronics and Information Technology.

The Conférence Européenne des Administrations des Postes et des Télécommunications (CEPT) is the association of European Postal and Public Telecommunications Operators and establishes standards and agreements relating to public telecommunications networks.

European Telecommunications Standards Institute (ETSI). At the suggestion of CEPT and with support from the European Commission, ETSI was established in 1988 to develop telecommunications standards for use in Europe. It has a similar scope to ITU-T and its membership is drawn from Public Telephone Operators, equipment manufacturers and users. An ETSI standard is called an ETS (European Telecommunication Standard).

The European Workshop for Open Systems (EWOS) was founded in 1987 with the support of the EEC and EFTA. EWOS develops OSI functional standards (profiles) and the corresponding test specifications. A functional standard specifies how to use other standards - base standards - to achieve a particular function. This approach is necessary because there is often a choice of base standards, each of which is capable of meeting the requirements of a particular application. Also, once a base standard has been selected, it is necessary to chose from the options that are contained within it. If this were not done, then it would be possible to have two pieces of equipment which, despite the fact that they both conformed to the same standard, would not interwork.

4.4.3 Standardisation in the United Kingdom

The principal standards authorities in the UK are the government's Department of Trade and Industry (DTI) and the British Standards Institute (BSI), an independent body incorporated under Royal Charter.

The DTI represents the UK as its Telecommunications Administration in such bodies as the ITU-T and ETSI. Its role is primarily policy and coordination and it runs committees such as TAPC (Telecommunications Attachments Policy Committee). The views of UK industry are obtained through *coordination committees,* with coordinators being appointed by the DTI. For example, the coordination committee for ITU-T Study Group 14 meets regularly to formulate the UK position on modem issues and discuss the progress being made in the Study Group.

The BSI is the UK representative in such bodies as ISO, IEC, CEN and CENELEC. Automation and Information Technology is the responsibility of DISC which is a self-financing division within BSI. It directs the work of four Standards Policy Committees, the two most relevant to Data Communications being Information Systems Technology (IST/-) and Telecommunications Equipment (TCT/-). Each of these has a group of Technical Committees entrusted with the production of UK standards and providing a UK view on regional and international standards. The Technical Committees set up formal sub-committees or panels of experts to work on particular projects. BSI IST/- parallels ISO/IEC JTC1.

Various UK manufacturers' associations take part in standardisation. A notable player in data communications is the Electronic Engineering Association (EEA).

4.4.4 Standardisation in the United States of America

The principal standards organisation in the USA is the American National Standards Institute (ANSI). It does not make its own standards (indeed its charter forbids this) but rather "accredits" other bodies to do so. ANSI is the US representative at ISO and IEC. The main organisations involved in telecommunications standardisation are listed below.

The Exchange Carriers Standards Association (ECSA) sponsors the T1 Telecommunications committee. This has a number of sub-committees including T1S1 which is concerned with 64 kbit/s and Broadband ISDN and T1X1 which studies the Digital Hierarchy and Synchronization.

The Computer and Business Equipment Manufacturers Association (CBEMA) sponsors X3 which covers all aspects of information systems. In particular X3S3 studies data communications (parallel to ISO/IEC JTC1 SC6).

The Telecommunications Industries Association (TIA) sponsors a number of committees such as TR29 (Facsimile Systems and Equipment) and TR30 (Data Transmission Systems ands Equipment).

Institute of Electrical and Electronic Engineers (IEEE) has a number of standards development projects. One such is 802 which is concerned with Local Area Networks (LANs).

4.5   Telecommunications standardisation in Europe - the role of the European Commission

The primary aim of the European Economic Community (EEC) is to create an open market in products and services. In the telecommunications market the aim is that the customer should have a choice of suppliers for both terminal equipment and telecommunications services and that these services should be available across national borders. Indeed, the Commission sees the liberalisation of telecommunications and the consequent reductions in prices from the resulting competition as important enabling factors in the free market and vital to the EEC being able to compete effectively with its North American and Far Eastern rivals. Since the degree of liberalisation varies enormously from one country to another, the Commission has issued a series of Directives which all countries must follow and which successively increase the openness of the telecommunications market.

4.5.1 Terminal equipment

Before an item of telecommunications terminal equipment is allowed to be connected to a public network it must be "approved". The Commission has issued a number of directives that cover these approvals.

4.5.1.1 First Phase Terminal Directive (86/361/EEC)
This was the first step on the path to Europe-wide approvals. At the request of the Commission, ETSI developed a series of standards, which when approved became NETs (Normes Européennes de Télécommunications). There are two kinds of NET. *Access NETs* are intended to ensure that no disturbance occurs to the network and that calls can be routed successfully. *Terminal NETs* ensure end-to-end compatibility of a communications service. Like many national approvals standards, NETs are "type approvals". An initial sample of a product is tested and, if approved, then all those subsequently manufactured are considered approved. If the product is modified, then in the case of a modular product, the modified parts must be approved to the current version of the standard. The unmodified parts need not be re-approved even if they were originally approved to an earlier version of the standard. To gain approval, a product has first to be granted a NET certificate by a test house (in any EEC or EFTA country) and then has to be granted a separate approvals certificate by the approvals body in every country in which it is to be sold. In general, NETs have a common part (which applies to all countries) together with a number of national annexes. In order to obtain approval in a particular country, it must have been tested to that country's annex.

4.5.1.2 Second Phase Terminal Directive (91/263/EEC)
The purpose of this directive is to harmonise between countries conditions for placing terminal equipment on the market. It requires that terminal equipment shall satisfy

certain "essential requirements" which include safety, network protection, electromagnetic compatibility (EMC) with the network, interworking with the public network for controlling connections and in specific ("justified") cases, interworking with other terminal equipment.

Harmonised standards called Common Technical Regulations (CTRs) give the specifications and tests which allow terminal equipment to be proved to comply with the essential requirements. (A harmonised standard is a Europe-wide one which replaces any conflicting national standard.) No national variations are permitted in CTRs.

The list of CTRs (with their numbers) includes:

1	X.21 network access
2	X.25 network access
3	Basic Rate ISDN access
4	Primary Rate ISDN access
8	Digital telephony over ISDN
11-18	ONP digital and analogue leased lines

The Second Phase Directive rescinds the First Phase Directive and this has resulted in some NETs being replaced by CTRs. In other cases where there are irreconcilable differences between countries this has not been possible and there are special arrangements to allow the use of NETs to continue. An example of this is NET 20 (modems) which calls up NET 4 (PSTN terminal equipment). NET 4 has a set of annexes which describe the national variations in the PSTN interface.

4.5.1.3 Production of CTRs

The production of a CTR is initiated by the Approvals Committee for Terminal Equipment (ACTE) of the Commission. This committee, with the advice of TRAC (Technical Recommendations Applications Committee), defines the requirement and scope statement for the new CTR. Based on the scope statement, the European Telecommunications Standards Institute (ETSI) prepares a work package which results in an ETSI deliverable, a TBR (Technical Basis for Regulation). A TBR can draw on ETSs (European Telecommunications Standards) and other base standards. When a TBR has been approved by ACTE it becomes a CTR.

There are a number of key dates in the life of a CTR once it has been adopted by the Commission. The first is its "date of effect". From this date it may be used in place of any existing standard. The next is the "end of the first transition". From this date all new products must be approved to the CTR although upgrades to existing products may continue to be approved to an older standard. Finally the "end of the second transition" is the date after which all products, both new and upgrades, must be approved to the CTR.

4.5.1.4 Gaining approval

An important point to note concerning CTRs is that the product is "approved to be placed on the market". This is unlike earlier European approvals (NETs) and national approvals which are "type approvals". With "approval to place on the market" a product must conform to the latest standard at the time it is sold. Thus, if either the product or the standard changes, the product must be re-approved. Equally, an upgrade must be approved to the latest standard.

To gain CTR approval, a product must be issued with an approval certificate from an authorised approval body. It may then be sold without further formality in any member state of the EEC or EFTA (European Free Trade Association). No further country-specific testing is required. There are three routes to gaining approval:

1. Testing and certification of a manufacturer-provided sample by a test house followed by further testing of samples purchased by the approval body on the open market.

2. As (1), but instead of the testing of purchased samples, the manufacturer is subject to a factory inspection and certification to ISO 9002 (BS 5750 part 2).

3. Internal testing by the manufacturer following inspection and certification to ISO 9001, (BS 5750 part 1 - design and manufacture). The manufacturer must also provide a Declaration of Conformity (to ISO 9001) to the approval body which administers ISO 9001 in that country.

4.5.1.5 Low Voltage Directive (LVD) (72/23/EEC)

This directive (LVD) describes the safety requirements for equipment "designed for use with a voltage rating of between 50 and 1000V AC and 75 and 1500V DC". The Second Phase Directive extends these requirements to all terminal equipment. The LVD covers not only electrical safety but also such things as acoustic shock in telephones, lightning protection and mechanical safety (eg. lack of sharp edges). The LVD allows manufacturers various ways of meeting safety requirements.

CTRs do not impose specific requirements for safety since terminal equipment must satisfy the requirements of the LVD. A CTR may, however, for information purposes, identify safety standards which would meet the requirements.

4.5.1.6 Electromagnetic Compatibility (EMC) Directive (89/336/EEC)

This directive (EMC Directive) requires electrical equipment to have appropriate RF emission and imunity characteristics and describes how a manufacturer can demonstrate conformity. A CTR must describe any additional specific requirements.

4.5.2 Services

One of the aims of the Commission is that common services should be available in all EEC countries. Another is there should be free competition in service provision. Two directives exist to further these aims.

4.5.2.1 Open Network Provision (ONP) Directive (90/387/EEC)

This covers the provision of standard network services across Europe, effectively removing boundaries between countries and making it as easy to rent an international service as one within the borders of one country. The customer will be able to enter into a contract with a single service provider, for a service extending between two or more countries. The interface in each country will be the same and the customer will have a single point of charging and fault reporting. This should be contrasted with present arrangements where, for example, a leased circuit customer has to sign a contract with the provider of each end of the circuit and has two charging and fault reporting points. Also the two ends may have different interface characteristics with different equipment approvals.

A number of "daughter" directives specify the individual services subject to ONP. They cover:

Digital leased lines at 64k and 2.048 Mbit/s
Analogue leased lines
Packet and circuit switched data services
Digital voice
Certain aspects of mobile (radio) services
Broadband ISDN (B-ISDN)

4.5.2.2 Services Directive (90/388/EEC)

This aims to control, i.e. reduce, the PTOs (Public Telecommunications Operators) monopolies in the provision of services. The Commission recognises that Europe is falling behind the USA and Japan in the the provision of telecommunications services and wishes to introduce more competition both within countries and across borders. For example, it should be possible for a customer who requires an ONP service between a given pair of countries to have a choice of supplier. A hole in the present proposals is that they exclude telephony over the PSTN (Public Switched Telephone Network) which at present comprises 90% of all telecommunications business. The Directive specifies that there should be a review of the telecommunications market in 1992 and 1994 with the possibility of further legislation.

4.5.3 Public Procurement Directive

The Commission has been concerned for some time that governmental and related organisations should benefit from the advantages of using standardised equipment (open systems). In order to give manufacturers an incentive to produce products conforming to OSI standards the Commission issued a directive to the effect that a "public body" must insist on conformance to OSI standards for any purchase exceeding 100,000 ECUs (European Currency Units). An exception to this rule is where an existing network is being upgraded and an OSI conformant upgrade would be incompatible with the rest of the network.

4.6 Developing a standard

4.6.1 Originating the idea

The need for a standard may be recognised in a number of ways:

1. It may arise from a planning exercise by one of the standards organisations.

2. It may result from a request from an industrial organisation for study in a particular area.

3. It may be identified by a national government or the European Commission as being needed as part of the approvals or regulatory process.

4. It may be the recognition of a *de facto* standard.

The actual process of producing a standard varies somewhat from one standards body to another and below are some examples.

4.6.2 Development of an ITU-T (formerly CCITT) recommendation

The work program of the ITU-T is specified in a series of Questions, a number of which are assigned to each Study Group. Each question is studied by a Rapporteur's Group of experts which meets sufficiently often to to progress the work (2-4 times per year). The text of each draft Recommendation is maintained by an Editor and is updated as a result of written contributions to the meetings and by discussion. When a draft Recommendation is considered complete it is offered for approval by the Study Group. If there is not total agreement then it is returned to the Rapporteur's Group for amendment. Once a draft Recommendation has been approved by the Study Group it is submitted for approval either by the next WTSC meeting (once every four years) or alternatively to an *accelerated procedure* which allows approval to be obtained in about seven months. Once approved the Recommendation is published by the ITU and may be purchased by any interested party.

 In the UK, the DTI-sponsored coordination committees meet to prepare UK contributions to meetings and to advise the DTI on how to vote.

4.6.3 Development of an ISO/IEC JTC1 Standard

There are five successive stages of technical work within JTC1:

1. A proposal for a *new work item* is circulated to members of JTC1. If this is approved then work can commence.

2. A *working draft* document is prepared within a Working Group.

3. When sufficiently advanced, the document is submitted as a *Draft Proposal* (DP) to members of JTC1 or one of its sub-committees for comment and vote.

4. When there is substantial support for the DP, possibly after revision, it goes forward as a *Draft International Standard* (DIS) for circulation to national standards bodies for approval.

5. When a DIS has received the support of the majority of JTC1 members and at least 75% of the national bodies, it is published as an International Standard (IS).

In addition, a *fast track* procedure exists to allow existing standards from other organisations to be submitted directly for voting as a DIS.

 In the UK the BSI IST committees, with input from sub-committees and panels of experts, prepare UK contributions and determine the national voting position.

4.7 Commercial versus technical considerations

Although a significant part of the international standards community consists of officals from governments and national standards bodies, many of the technical experts in the working groups are from industrial organisations. This has considerable advantages both for the standards committee, in that expertise and resources are available to progress the work, and for the manufacturers in that they are able to influence the direction of development of a standard.

 One aspect of standards development that may result in conflict between technical and commercial interests relates to intellectual property. If there are alternative patented solutions to a problem then the rival patent holders will each try to ensure that their solution is incorporated in the standard since this is likely to result in substantial licence revenue. Both the ITU-T and ISO/IEC JTC1 require that companies holding patents which may cover some aspect of a standard make a statement concerning the availability of licences. Failure to do so is likely to result in the standard being withdrawn. The usual statement is that the company is prepared to make licences available on reasonable and non-discriminatory terms.

4.8 Maintenance of standards

No approved and published standard is perfect, however much effort and time have been expended in its creation. Experience in using a standard will show up its defects and in addition, as time goes on, various possible enhancements will become apparent. These may result from changing needs or advances in technology. Thus, inevitably, alterations will be proposed. Where a change is necessary to correct an error which prevents the correct operation of a piece of equipment it is obvious that the change should be published without delay. In this case the standards organisation can publish a supplement to the standard containing the correction.

An enhancement to a standard is a more difficult matter and presents a number of problems. One is, quite simply, that the development time for a piece of communications equipment is measured in years rather than months and changing the specification during development can only increase costs and delay completion. Another, perhaps more serious, is where a piece of equipment built to the enhanced standard would be incompatible with one built to the old standard. To minimise these problems, standards are designed, as far as is reasonable, to be extendable and such that a piece of equipment conforming to the newer version can "recognise" one conforming to the older and downgrade its operation. Also, to avoid frequent changes, proposed enhancements will often be accumulated by a standards organisation and the new version of the standard will be issued several years after the old one.

Corrections and changes to a standard can be proposed by a member of the standards organisation, or sometimes thorough a liaison from another standards organisation. In the ITU-T, a member will submit a written contribution to a Study Group meeting proposing the change. If the Study Group agrees that the Recommendation should be changed then the person responsible for editing the Recommendation (the editor) will produce a *Draft Revised Recommendation*. This then has to proceed through the same development process as for a new Recommendation. In the ISO, every standard is reviewed at least every five years by the committee responsible for it. In addition any national body can request a review at any time. If a decision is made to revise the standard then the revision follows the normal standards development process. An enhancement to a standard can be published as an Amendment, the development of which follows the same procedure as for a new standard.

4.9 How to obtain standards

The primary source of a published standard is obviously the standards body concerned. Each standards body also publishes a catalogue, listing all the standards currently available. Catalogues can be obtained from the relevant standards bodies or from the other sources mentioned below. For example, ITU-T Recommendations can be obtained from ITU headquarters in Geneva. BSI Publications, as might be expected, stock Bristish Standards and, as the UK representative, those of ISO and IEC. They also stock European standards from CEN, CENELEC and ETSI and Recommendations from the ITU. BSI will obtain standards from other regional and national standards organisations on request. (BSI Publications - telephone 0908 221166.)

Many university libraries have collections of standards which can be consulted and particular standards can be obtained through a local public library. Also, the BSI has a library which can provide advice and assistance. (BSI Library - telephone 0908 226888.)

Alternatively there exist a number of commercial organisations which can supply copies of standards. They are often quicker than BSI Publications, particularly for foreign material, but can be considerably more expensive. Sometimes draft standards and working documents can be provided, subject to copyright restrictions.

It should be noted that standards, like all written material, are subject to the copyright laws and may not be reproduced without the consent of the copyright owner.

4.10   Data communications standards in action

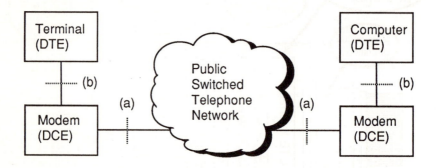

Fig. 4.3 *Data transmission interfaces*

Fig. 4.3 shows some of the interfaces involved when a terminal is connected to a computer via dial-up modems and the public telephone network. The interfaces of interest to the user are between -

    (a)   the modem and the telephone network.

    (b)   the terminal (or computer) and the modem.

Standardisation of interface (a) ensures that the modem works correctly with the network, does not interfere with other users and cannot damage the network. A standard is also necessary at this point if there is to be a free market in modems, i.e. the user is not forced to buy the modem from the telephone service provider.

Standardisation of interface (b) means that the user is free to buy the terminal and modem from different manufacturers and be confident that they will work together. Further standards may be required for the terminal and computer if they come from different manufacturers although these should not affect the modems.

Standardisation of functions within the modems ensures that they will interwork and that the desired end to end performance is achieved. Conformance with these standards can, of course, be verified only by tests between interfaces since there is no additional access to the modems' internal functions.

4.10.1   V32 modem

As an example of data communications standards in action, Fig. 4.4 shows the standards that apply to a typical V32 modem with data compression.

From left (user interface) to right (telephone network interface) the modem must comply with the standards listed below. The one concerned with the network interface is mandatory in that conformance is necessary to gain attachment approval. Others, although voluntary, are necessary in that these are what customers will be looking for in order to guarantee compatibility with other manufacturers' products. In addition, the modem has to conform to a number of other mandatory standards relating to electrical safety, network protection, electromagnetic radiation, flammability and component standards.

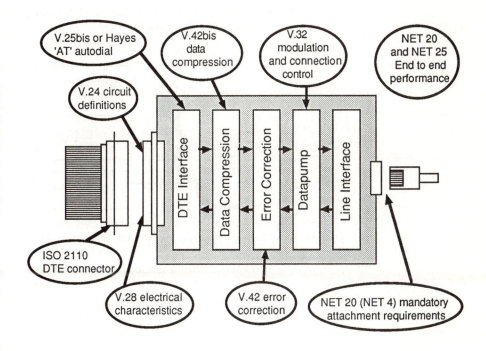

Fig. 4.4 *Standards in action - a V.32 modem*

ISO 2110 "25-pin DTE/DCE interface connector and pin assignments". This standard specifies the physical dimensions and the pin assignments for the 25-way "D" type connector.

ITU-T (CCITT) V.24 "List of definitions for interchange circuits between data terminal equipment (DTE) and data circuit terminating equipment (DCE)". This lists the interface circuits and specifies their function.

ITU-T (CCITT) V.28 "Electrical characteristics of unbalanced double current interchange circuits". This specifies the electrical characteristics of the interface circuits.

ITU-T (CCITT) V.25bis "Automatic calling and/or answering equipment on the general switched telephone network (GSTN) using 100 series interchange circuits". This specifies the protocol used to communicate with the modem for the purpose of automatic dialing.

A more comprehensive alternative is:
Hayes Standard "AT" Command Set. This is a *de facto* standard originally produced by Hayes Microcomputer Products Inc. It specifies a set of commands for the control of the modem through an asynchronous port. It is an example of a de facto standard which is in the process of being turned into a formal standard by the TIA (TIA-602) and the ITU-T (draft Recommendation V.at).

ITU-T (CCITT) V.42 "Error-correcting procedures for DCEs using asynchronous-to-synchronous conversion". This specifies the error control protocol used between a pair of modems. In fact, two error correction protocols are specified in the Recommendation. The main body of the Recommendation describes LAP-M, an HDLC-based protocol. (HDLC is an ISO standard.) Annex A contains an alternative protocol, the previously de facto standard, MNP4. A modem conforming to V.42 must implement LAP-M and may optionally implement MNP4.

ITU-T (CCITT) V.42bis "Data compression procedures for data circuit terminating equipment (DCE) using error correcting procedures". This Recommendation describes a protocol that runs over V.42. It uses an adaptive coding technique that allows a smaller number of bits to be sent over the line than are received over the terminal interface. This allows the terminal to operate at a higher bit rate than would otherwise be possible. For example, a 9.6 kbit/s modem could support a terminal operating at 19.2 kbit/s.

ITU-T (CCITT) V.32 "9600 bit/s duplex modem standardized for use on the general switched telephone network and on leased telephone type circuits". This defines the modulation scheme used by the modem and the means for establishing a connection.

ETS 300 114 (NET 20) "Attachment to the PSTN - Basic attachment requirements for modems standardized for use on the PSTN". The first part of this standard describes the category I (mandatory) requirements for the approval of the interface between the modem and the public telephone network. It refers to NET 4 (see below). The second part describes the category II (voluntary) requirements for end-to-end operation of a pair of modems that are common to all types of modem.

ETS 300 001 (NET 4) "General technical requirements for equipment connected to an analogue subscriber interface in the PSTN". The PSTN interface varies from country to country and so NET 4 has a set of national annexes, one for each country. The modem must be tested to the national annexes corresponding to each country in which it is to be sold. This means that there will have to be several different versions of the product.

ETS 300 002 (NET 25) "Approval requirements for 9600 or 4800 bits per second duplex modems standardized for use on the PSTN". This describes the specific category II (voluntary) requirements for end-to-end operation of a pair of V.32 modems.

4.11 Conclusion

Standards are essential to all players in the data communications arena. For manufacturers, an understanding of the standards development process (or even an involvement in it, perhaps via a Trade Association) is essential since they must ensure that:

- they understand to which standards the potential users expect the equipment to conform.

- they understand which standards are mandatory in order to gain approval to put the product on the market.

- the standards being developed are commercially viable and technically feasible.

The same applies to users (who may be involved either directly or through a Users' Group) because:

- they need to be aware of standards being developed which may affect their system and network purchases.

- they can ensure that standards being developed meet their needs.

- they can initiate standardisation activity in areas of concern.

- they can insist on conformance to established standards.

- they need to have confidence in the ability of their data communications to migrate to Open Systems (OSI).

4.12 Acknowledgements

The author wishes to thank the following for their helpful advice and assistance:

Dr. A. D. Clark, Hayes Microcomputer Products Inc.
Mr. M. Harris, Cray Communications Ltd.
Mr. R. Walker, DISC/BSI Standards.

Chapter 5

ISDN

Ron Haslam

5.1 Introduction

In 1992 there were 960,000 channels of access to Integrated Services Digital Networks (ISDNs) in Western Europe. It has been estimated that within ten years twenty eight million channels, or one half of all business exchange lines in Europe, will be connected to an ISDN. In December 1993 the European Commission produced a white paper on the European economy [1]; the paper recommended the building of a Europe-wide ISDN initiative at a cost of ECU 15 billion, targeted at the medium and small business market, and residential users. However the main driving force behind ISDN has been technological. ISDN is the natural progression resulting from the modernisation of the Public Switched Telephone Networks (PSTN) from an analogue network to a more flexible and 'future proofed' digital network. Most major network providers are now looking towards ISDN as being the only way to access their networks in the future, a single interface offering the user combined voice, data, text, video and still image applications, in a simple, convenient, pre-provisioned and user-controlled manner.

In the United Kingdom modernisation started with the introduction of Pulse Code Modulation (PCM) into the main trunk network and the replacement of analogue switching with digital switching to form an Integrated Digital Network (IDN). The final stage to reach an end to end digital system was the extension of the digital capability into the local loop all the way to the user. All that remained then for the launch of an ISDN was to ensure that ISDN terminals were available, viable in cost and application, and capable of interworking both nationally and internationally.

5.2 ISDN concepts

The opening paragraph of the International Telegraph and Telephone Consultative Committee (CCITT) I series Recommendations on ISDN states:

" An ISDN is a network in general evolving from a telephony IDN that provides end to end digital connectivity to support a wide range of services, including voice and non-voice services to which users have access by a limited set of standard multi-purpose user-network interfaces "

This was not a paragraph produced by a flash of inspiration but a carefully planned stage in a universal move to bring telecommunications in line with the communications requirements of a rapidly advancing society based on high speed technology.

It has been a long struggle to reach '*an ISDN*'. Many attempts to transfer information quickly over long distances had been tried in the past; flag waving, smoke signals, bonfires and tom-toms are the more well known examples. By 1875 most Western countries had set up telegraph systems using two wires and at about that time the first telephone arrived allowing direct speech. By 1936 the first telex network had appeared in Germany. The public telegraph, telephony and telex networks had now started to form as three totally separate telecommunications networks

The advent and rapid spread of the transistorised computer in the early 1950s meant a sudden demand had to be met for the transfer of large amounts of data. Of the three main networks the telephony network was chosen to carry the data, perhaps because of its widespread user base. Limitations of data transfer over the analogue network and problems of call set-up times on electromechanical switches forced a decision in the 1970s to set up two specialised networks for inter-computer working; in Europe these were the packet network (as current in the UK) and a 'fast call set-up' circuit switched network operating at kilobits per second (not in the UK). At this time it was also recognised that the telephony network itself was rapidly becoming 'old technology' and needing modernisation. Network operators needed an answer, something that met the increasing demands for faster data transfer yet still catered for 'traditional' speech circuits, but at the same time was cost effective and 'future proofed'.

The solution lay back in the 1930s when a process known as Pulse Code Modulation (PCM) was conceived. PCM is a process of sampling analogue signals 8000 times a second and encoding the signals into one of 256 possible levels represented by 8 binary digits, the analogue signal therefore being transmitted in an encoded binary stream of 64 kbit/s. The drawback in the 1930s to the development of a PCM system was that the process was forty years ahead of the technology needed to realise a viable telecommunications network. Developments in fibre optics and Very Large Scale Integrated Circuits (VLSI) in the 1970s reversed the situation. By the 1980s all the major network operators had completed a modernisation programme of their Public Switched Telephone Networks (PSTNs), each operator creating a totally digital trunk network, an Integrated Digital Network (IDN).

IDNs use digital connections between exchanges, structured on the 64 kbit/s PCM and using computer controlled call set-up via separate links (common channel signalling).

The user's concept of ISDN is of a single network access, 'the hole in the wall', through which the user has access to all the services available when needed. It is difficult for the user to justify in cost, or efficiency, the use of dedicated circuits provided on a permanent basis unless they are being used for transmission of a continuous stream of data, or a vast amount of information transfer. Most users only require intermittent transmissions so the idle periods prove costly. The network operator offering access to all services via a single interface means that the user requires only a single dedicated circuit to the local exchange where the switching function is performed. Sharing all the services over that one access means that the user pays call charges only for the time of actual use of the services.

5.3 The current ISDN situation

Recommendations made internationally by the CCITT refer to the access (from the user to network interface) in terms of channels. The channels are provided over two main types of service, either the basic rate service which offers users a single connection of 144 kbit/s providing two 'B'-channels and a 'D'-channel, or the primary rate service which offers a user a single connection of 30 'B'-channels and a 'D'-channel (on 2 Mbit/s networks, e.g. UK), or 23 'B'-channels and a 'D' channel (on 1.5 Mbit/s networks, e.g. USA and Japan). The 'B'-channels are 64 kbit/s channels used for carrying user information such as voice, data, text, and video. The 'D'-channels, either 16 kbit/s for basic access or 64 kbit/s for primary rate access, are intended for carrying signalling information for circuit switching but may also carry packet data. Access recommendations for the ISDN are defined in CCITT Recommendation I.412.

5.4 Open System Interconnect (OSI) and ISDN

One of the major advantages ISDN has over its predecessor, the PSTN, is that signalling information is separated from the user data information. In CCITT terms the user data is in the user 'plane' ('B'-channel) and the signalling is in the control 'plane' ('D'-channel). The signalling protocols in the control 'plane' are organised closely in line with the first three layers of the International Standards Organisation's (ISO) seven-layer reference model for OSI.

Fig. 5.1 shows a simplified diagram of how the control plane and user plane protocols relate to the seven-layer model. It is a very simplified diagram but illustrates message flow, the lines passing through the stacks showing the passage of messages through the stacks. In the user plane, information flow carries applications from terminal to terminal (layer 7 to layer 7). Since this is an unrestricted 64 kbit/s channel the user is free to use whatever protocol is required at layers 2 and 3 and above. If the channel is to be used for data transfer then any protocols relevant to the users data terminal would be used, e.g. SDLC, HDLC, etc.

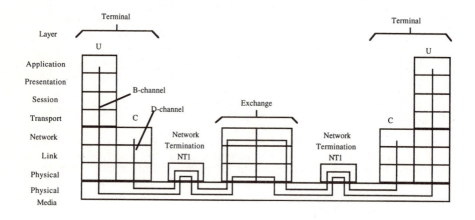

Fig. 5.1 *The 7-layer model in relation to ISDN channels*

In the control plane the information carried in the 'D'-channel is in the form of signalling information and is clearly defined in the first three layers of the model.

5.4.1 Interface recommendations in relation to the layer model

Tables 5.1 and 5.2 show the relationship between the first three layers of the layer model and the relative CCITT recommendations for ISDN. CCITT recommendation I420 refers to all the I series of recommendations for basic rate. Similarly the I.421 recommendation covers the primary rate series of recommendations.

LAYER 3	SUPPLEMENTARY SERVICES	I.452/Q.932	
"	SPECIFICATION	I.451/Q.931	
"	GENERAL ASPECTS	I.450/Q.930	I.420
LAYER 2	SPECIFICATION	I.441/Q.921	
"	GENERAL ASPECTS	I.440/Q.920	
LAYER 1		I.430	

Table 5.1 *CCITT basic rate interface recommendations*

LAYER 3	SUPPLEMENTARY SERVICES	I.452/Q.932	
"	SPECIFICATION	I.451/Q.931	
"	GENERAL ASPECTS	I.450/Q.930	I 421
LAYER 2	SPECIFICATION	I.441/Q.921	
"	GENERAL ASPECTS	I.440/Q.920	
LAYER 1		I.431	

Table 5.2 *CCITT primary rate interface recommendations*

Tables 5.1 and 5.2 show that the two sets of recommendations only differ at layer 1.

5.5 The ISDN reference configuration

The CCITT have defined a number of ISDN reference points where equipment interfaces may exist in various functional configurations (Fig. 5.2).

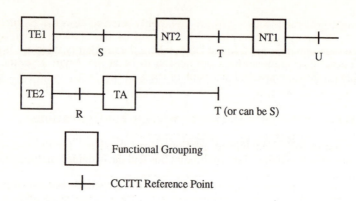

Fig. 5.2 *ISDN user-network reference configuration*

Where:

NT1   Network Termination 1 - This terminates the two-wire line from the network and on the user side provides the 4-wire interface. Provides broadly the layer 1 functions of the OSI model, i.e. correct physical and electrical termination of the network by providing line transmission termination, and for access to the network: timing, power feeding, multiplexing and contention resolution .

NT2   Network Termination 2 - This provides functions to interface private networks to the ISDN access, examples would be a Private Automatic Branch Exchange (PABX) or LAN Bridge or Router. Functions include those associated with the first three layers of the OSI model, i.e. physical, data and network layers.

TE1   Terminal Equipment 1 - Includes functions associated with any terminal equipment that complies with the ISDN interface specifications, examples being digital telephones, group 4 fax and video phones.

TE2   Terminal Equipment 2 - A terminal not complying with the ISDN specifications but has an interface complying with another network such as analogue or data networks, e.g. CCITT recommendations such as CCITT V-series for modems or X-series for data networks.

TA Terminal Adapter - Includes the functions for adapting any equipment that does not meet ISDN interface requirements such as converting V.24 or X.21 to the ISDN 'S' or 'T'-interface.

The connection from the network to the user's premises is provided over the standard 2-wire copper loop; in CCITT terms the 'U' reference point. Between the network termination NT1 and the terminal, transmission is carried over a balanced four wire bus (Fig. 5.5).

Note that the 'R', 'S', and 'T' reference points represent the physical interfaces that are defined in ISDN standards but can actually be combined in a single item of equipment. The NT2 may be omitted, in which case the 'S' and 'T' reference points are the same point. Although the 'S' and 'T' are strictly speaking two reference points at which interfaces may occur, they are commonly referred to as the 'S' interface. In the UK, as in most countries, the NT1 is owned by the network provider and is located on the user's premises. In the USA, where the regulatory interface is decreed by the Federal Communications Commission to be at the 2-wire line, the user is responsible for the purchase and provision of the NT1 functions.

5.6 Basic rate user-network wiring configurations

The most commonly used wiring configurations specified for the user-network interface are: point-to-point, short multipoint bus and the extended multipoint bus (see Fig. 5.3).

The point to point configuration can have only one terminal connected at the end of the 4-wire bus. The maximum length of the bus is 1 km, cable attenuation is 6 dB at 96 kHz. The short multipoint bus has a maximum length of 200 m and is designed to allow up to eight terminals to be connected at points anywhere along its length. By using an extended bus, terminals can be situated at the end of a longer bus, the maximum length should be 500 m with a maximum distance between terminals of 25 - 50 m.

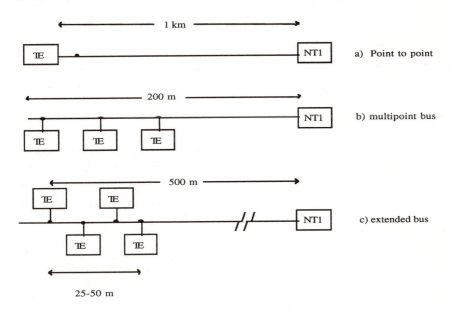

Fig. 5.3 *User-network interface bus configurations*

5.7 ISDN services

In the context of ISDN, services can be divided into two broad categories: bearer services and teleservices.

5.7.1 Bearer service

This is the service the user accesses when connecting to an ISDN interface, i.e. the bearer service provides the capability for the transmission of signals between user-network interfaces and involves only the lower three layers of the 7-layer model. There are three types of ISDN switched circuit bearer services available in the UK:

1   *64 kbit/s unrestricted*
An unrestricted bit-transparent path, allows data transfer at up to 64 kbit/s. Data rates below 64 kbit/s are catered for by rate adaptation techniques. Call progress tones and announcements are not provided (call progress information is provided via 'D'-channel signalling), and no interworking with the PSTN is allowed.

2   *3.1 kHz audio*
This offers the ISDN user similar features to the PSTN, allowing the transfer of audio information from modems, group 3 fax machines, etc. Echo control and Digital Circuit Multiplication Equipment (DCME) are not used. (On the PSTN, disabling tones of 2100 Hz from a terminal control these devices on the networks).

3   *Speech*
Offers all the normal routing rules for speech (i.e. Echo control may be required, DCME can be used on international routes, no more than two satellite links and A-law to µ-law conversion required when operating between countries using different coding techniques).

5.7.2 Teleservices

These provide the complete service capability including the terminal and network functions, allowing communication to take place between users according to agreed protocols usually operating over one of the bearer services, i.e. a teleservice generally involves both lower (layers 1-3) and higher layer functions (layers 4-7).

A user selects the service to be used at the time of call establishment. Teleservices have a very precise definition of terminal requirements and, since the UK is fully liberalised, terminals can be purchased on the open market and, provided they have a conformance to standards approval, can be connected to the network. Therefore BT, for example, does not always offer complete teleservices in the UK but does provide the necessary network features for users to use a teleservice. As shown in Fig. 5.4, teleservices are accessed via terminal equipment. Bearer services can be accessed at either the 'S' or the 'T' interface depending upon what equipment the user has chosen.

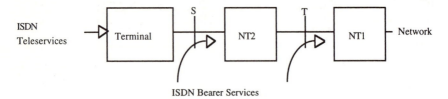

Fig. 5.4 *Access to the ISDN services*

A third 'service' is available known as 'supplementary services'. Supplementary services are options that are not stand-alone functions but are available with both bearer services and teleservices. They are regarded as modifications or supplements to a basic telecommunication service, e.g. CLIP, and three party calling.

5.8 User-network interface

5.8.1 Layer 1 (The Physical Layer)

The physical layer covers the physical transmission medium connecting the user to the network (wires). Basic and primary rate protocols for access interfaces are also specified in this layer. It is the lowest layer of functionality in an ISDN and deals with transfer of information across the interface from terminal to network and from network to terminal.

5.8.2 The functions of basic rate layer 1

In order to provide a transport service, layer 1 must perform the following functions: power feeding, 'B' and 'D'-channel management, 'D'-channel access contention, deactivation, frame structure, frame alignment, bit and octet timing, mechanical interface, and bit transmission.

5.8.3 Electrical characteristics

The electrical characteristics of the user-network interface are specified in CCITT Recommendation I.430. The characteristics cover jitter, pulse shaping, delay characteristics, bit rate, input and output impedances for both the terminal and the network termination receiver sensitivity, and power transfer across the interface. Fig. 5.5 shows the electrical interface for the user-network interface.

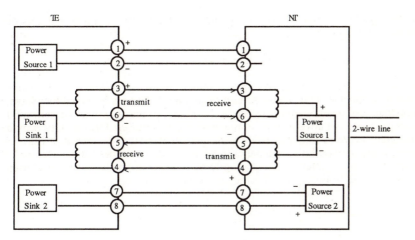

Fig. 5.5 *The electrical interface of the basic rate user-network interface*

Wires 3, 6, 5 and 4 are mandatory; they are used for the transfer of information across the interface. The polarity of the framing pulses is shown on the transmit and receive wires. Wires 1, 2, 7 and 8 are optional wires provided for alternative power feeding.

Connections are made into ISDN via 8-pin plugs and sockets made to the International Standards Organisation (ISO) formal specification ISO 8877.

5.8.4 Power feeding options

With an ISDN there is no specific requirement for an administrator to supply power over the network interface. Power for the terminals is normally provided for by local mains power via one of three ways, power source 1, power source 2 (both via the NT1) or power source 3 (part of the terminal).

5.8.5 Power source 1

Provided in the NT1 is a supply, via the centre tapped transformers (Fig. 5.5), of at least 1 Watt at 40 volts (- 15%, +5%), but, in the event of local power failure, power will be fed via the network and restricted to a minimum of 420 mW. The actual power available for a terminal to draw under these conditions is 380 mW while operating and 25 mW when powered down. To indicate that local power has failed to all terminals connected to the NT1, the power feed polarity will go in reverse to that shown in Fig. 5.5 (normal power condition). This restricted power is intended to drive a single digital phone for emergency use.

5.8.6 Power source 2

Provides a maximum of 7 Watts at 40 volts derived from the local power supply .

5.8.7 Power source 3

This source is provided by the terminal and, as such, is not subject to CCITT Recommendations.

5.8.8 Deactivation of the NT1 and TEs

In the UK, the NT1 and TE are placed into a low-powered state (deactivated) when no calls are in progress. This is to conserve power in the network (NB. unlike the basic rate interface, the primary rate interface is considered to be continuously active and no start-up procedure is defined). The saving with deactivation of the basic rate interface by the operator is considerable since the NT1s are network-powered. Secondly, electromagnetic radiation is reduced and hence crosstalk interference reduced. The network interface may be de-activated only by the network because of the multi-bus function of the NT1, but activated by either the network or the terminal.

5.8.9 Basic access frame structure

The basic access frame on the user-network interface is shown in Fig. 5.6. Each frame lasts for 250 s and contains 48 bits of 5.2 s duration. The transmission rate is therefore 192 kbit/s. The frame format is the result of the time division multiplexing of the two 64 kbit/s 'B'-channels and one 16 kbit/s 'D'-channel. It is made up of two octets (16 bits) of channel B1, two octets of channel B2 and four bits of the 'D'-channel. The remaining (12) bits are used for framing, balancing and echoing. Balance bits (L bits) are inserted to prevent a build up of dc on the line. The terminal (TE) timing is derived from the frames transmitted by the NT1.

Each frame from the TE is delayed by two bits with respect to the frame received from the NT1. The line code used is Alternate Mark Inversion (AMI). In this code a logical 1 is transmitted as a null, or zero voltage, and a logical 0 as a positive or negative voltage with a nominal pulse amplitude of 750 mV.

5.8.10 Shared 'D'-channel access and contention

Any terminal wishing to access the 'D'-channel has to follow a number of steps. These include: basic signalling requirements, contention procedures and a priority scheme.

The basic signalling requirements are that: the terminal transmits continuous binary ones (represented by zero volts) when idling; the terminal physical layer is in frame and bit synchronisation with the NT1; and the terminal monitors all 'D' bits echoed back by the NT1 ('E' bits) comparing them against the 'D' bits it transmitted. Priorities are set in relation to the number of consecutive one bits which must be counted before the terminal can access the 'D'-channel; the lower the number of bits, the higher the priority. When a terminal has just completed a frame, the value of the count is incremented by one. This gives other terminals a chance at accessing the channel, allowing an access and priority mechanism to be established.

The 'D'-channel contention procedure ensures that when two or more terminals simultaneously attempt to access the 'D'-channel only one will be successful in transmitting information. This is carried out by each terminal comparing its last transmitted 'D' bit with the next 'E' bit received. If they are the same the terminal continues to send, if different the terminal detecting the difference will cease transmissions immediately, leaving another terminal to transmit.

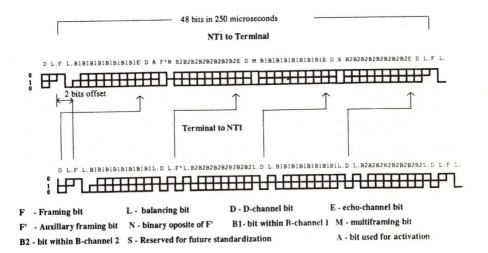

Fig. 5.6 *Basic access frame structure*

5.8.11 Primary rate interface

The primary rate interface is configured for point to point only, and is continuously active so no start-up procedure is defined. The electrical characteristics are defined in CCITT G.703. In the 2048 kbit/s primary rate interface the frame structure is based on time division multiplexing of 64 kbit/s channels. Each frame consists of 256 bits divided into 32 time slots, each slot containing an 8 bit sample from one channel. There are no framing bits, framing information is carried in slot 0 of every other frame (other information such as alarms are carried in the remaning frames). The repetition rate is 8000 frames per second which gives 2048 kbit/s. When 'D'-channel signalling is offered, the signals are carried in slot 16.

5.8.12 Basic rate terminal start-up

When a basic rate terminal wishes to start-up, i.e. move from powered-down state to fully activated state, it follows a procedure which is state-driven as shown in Fig. 5.7.

An inactive terminal emits no signal to line. This signal is called Info 0. When terminals request activation they begin to transmit continuous Info 1 signals. Info 1 is an 8-bit pattern of 00111111. (Note this signal is not synchronised to the network).

The NT1 will respond to an Info 1 with an Info 2 signal. This signal is a formatted frame with binary zeros in the data channels (i.e. all B, D and E bits set to zero). The high density of binary zeros permits terminals to synchronise very quickly.

Terminals, when synchronised, will respond with Info 3 frames containing operational data. The NT1 will return Info 4 frames.

Since all terminals are connected to the bus in parallel and are therefore activated in parallel, it is not possible to activate just one terminal on a multi-terminal bus.

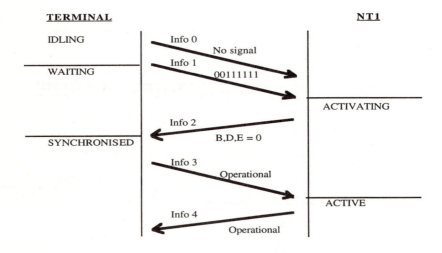

Fig. 5.7 *Activation/de-activation procedure*

5.8.13 Basic rate layer 2 (the Data Link Layer)

The Data Link Layer in ISDN describes the High Level Data Link (HDLC) procedures. The process is more commonly referred to as the Link Access Procedure on the 'D'-channel or by the acronym LAP D. Layer 2 provides all the protocols and services necessary to provide a secure error-free connection between two end points connected by a physical medium. LAP D is based on LAP B of the X.25 Layer 2 recommendation, but in ISDN the Data Link Protocols associated with the 'B'-channel are not defined by any standard since the 'B'-channel is offered to the user as a clear 64 kbit/s channel and it is up to the user to use any protocol considered suitable.

LAP D offers frame multiplexing by having separate addresses at layer 2, allowing many LAPs to exist on the same physical connection. The ability to have many LAPs on the same physical connection offers the possibility of many terminals being connected to the same line. Internationally, it has been agreed that this will be fixed at a maximum of eight terminals but, theoretically, the bus could support many more.

The layer 2 frames consist of a sequence of consecutive 8-bit elements as shown in Fig. 5.8. A fixed pattern of 8-bits called a flag is provided at the start and also at the end of each frame. The second and third octets are the LAP identity or LAP address field. The address consists of a Service Access Point Identifier (SAPI), Terminal Endpoint Identifier (TEI) and a command/response bit.

octet 1	0 1 1 1 1 1 1 0	Opening Flag
octet 2	address octet 1	
octet 3	address octet 2	
octet 4	control octet 1	Control field structure depends on the frame type.
octet 5	control octet 2	The second octet in the control field is not always present.
octet 6	layer 3 information	Layer 3 information only present in layer 2 'information frames'.
octet n-2	FCS octet 1	Frame check sequence.
octet n-1	FCS octet 2	Frame check sequence.
octet n	0 1 1 1 1 1 1 0	Closing flag.

Fig. 5.8 *Layer 2 frame structure*

5.8.14 The address field

The address is only of significance to the local two end-to-end points on the user interface using the LAP.

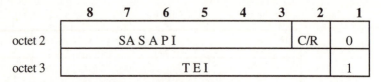

	8	7	6	5	4	3	2	1
octet 2			S A S A P I				C/R	0
octet 3				T E I				1

Where: SAPI = Service Access Point Identifier
        TEI = Terminal Endpoint Identifier
        C/R = Command/response bit

Fig. 5.9 *The address field*

The LAPD address field is shown in Fig. 5.9. Bit 1 of each octet is used as the extension indicator in compliance with HDLC rules, i.e. the first bit of octet 2 is set to zero to indicate additional octets and the first bit of octet 3 is set to one to indicate the last octet of the address. Bit 2 of octet 2 is a command response indicator; this is used in conjunction with the address field, by a terminal/NT1, to separate incoming frames into commands that require processing by the receiver and the responses that require processing action by the transmitter. This is used in ISDN since there is no master/slave relationship between a terminal and the NT1, both being able to initiate actions at will. The simple rule is that commands carry the address of the distant termination (TE or NT1) while responses carry the address of the source termination (TE or NT1). In LAPD the user's terminal always sends commands with C/R set to zero and responses with the C/R bit set to one. The network side (i.e. NT1) uses the opposite rule.

The seven remaining bits of the 3rd octet are the terminal endpoint identifier (TEI). This identifies a particular terminal associated with a particular SAPI. Each terminal connected to the user interface has a different TEI; the combination of SAPI and TEI will identify the LAP and will therefore form a unique layer 2 address.

All terminals transmit the layer 2 address in frames and only frames carrying the correct address are processed. It is therefore important that no two TEIs are the same. TEI values of 0-63 are used where non-automatic assignment is required, i.e. the value chosen is the responsibility of the user. Values 64-126 are automatically assigned values, the value being selected by the network. Value 127 (all ones) is assigned as the global or broadcast value. This permits broadcast information to all terminals within a given SAPI, e.g. a broadcast message to all faxes on a single interface offering an incoming fax..

Terminals having non-automatic values need not negotiate with the network before establishing a layer 2 connection. This procedure helps with terminal portability (the terminal could be moved to another location).

5.8.15 Control field

The control field (Fig. 5.10) is of one or two octets, depending on the frame type, and carries information relating to layer 2 sequence numbers used for link control. On long propagation delay links, such as satellite links, it may be necessary to extend the modulo 8 sequence count of basic mode control, i.e. the number of frames which can be transmitted before a frame can be received. The control field allows this capability. Extension by the addition of a second contiguous octet immediately following the basic field increases the modulo count to 128. In this extended mode, the transmitting terminal sets the P/F bits in octet 4 (bit 5) and octet 5 (bit 1). A receiver in basic mode, on receiving an extended mode control field, interprets the P/F in octet 4 (bit 5). A terminal already in extended mode interprets the P/F in octet 5 (bit 1),

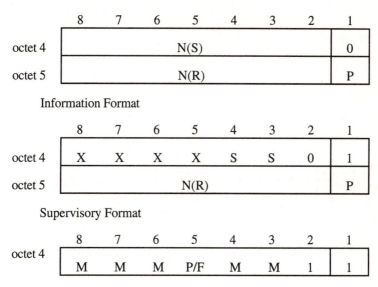

Information Format

Supervisory Format

Unnumbered Frame Format

N(S)   Transmitting send sequence number
N(R)   Transmitting receive sequence number
S   Supervisory function bit
M   Modifier function bit
P/F   Poll bit when issued as a command, final bit when issued as a response
X   Reserved bit, set to zero

Fig. 5.10   *Information, supervisory and unnumbered frame format*

5.8.16 The information field

The protocol is completely transparent to the information used in this field, the only requirement being that the information field must be an integral number of octets and a maximum of 260 octets.

5.8.17 Frame Check Sequence

Each frame contains a 16-bit Frame Check Sequence (FCS) to detect errors. If frames are corrupted for any reason at layer 1, then the frames will be received with invalid FCS values and will be discarded. The FCS is generated by dividing all the binary bits (excluding inserted zeros) in a frame, including the address, control, and information fields, up to the FCS by the generator polynomial (CCITT Recommendation V.41) $X^{16} + X^{12} + X^5 + 1$. The remainder forms the FCS. On receipt, the complete frame, including the FCS, is divided by the generator polynomial and, in the absence of transmission errors, the remainder should always be zero. Any other value indicates an error in transmission.

5.8.18 *Layer 3 (The Network Layer)*

This layer is concerned with the establishment and control of connections and the control of the supplementary services. Layer 3 call control information is carried in the information elements of layer 2 frames.

	8	7	6	5	4	3	2	1
octet 1	0	0	0	0	1	0	0	0
				Protocol discriminator				
octet 2	0	0	0	0	length of call reference value (in octets)			
octet 3	0	call reference type						
octet 4	0	message type						
	other information elements as needed							

Fig. 5.11 *Layer 3, signalling message structure*

There is no simple way of showing a message format since layer 3 messages contain many variables. CCITT I.451 numbered the octets as shown in Fig. 5.11, but this will vary between message type. The first four octets consist of:

Octet 1 - This octet contains a protocol discriminator which gives the 'D'-channel the capacity of simultaneously supporting other message types yet to be defined. The bits shown are the standard for user-network call control messages

Octet 2 - Bits 1-4 give the length of the call reference in octets, bits 5-8 are not used and are set to zero.

Octet 3 - The call reference value in bits 1-7 is used to identify the call to which a particular message is associated. It provides a reference for a set of messages associated with one operation or interaction between the terminal and network. Each call uses a unique randomly generated number for each operation. Bit 8 is a FLAG set to zero [0] by the originator or a one [1] by the destination terminal.

Octet 4 - Bits 1-7 identify the message type, i.e. a CONNect message, or a SETUP message. Bits 6 and 7 group the messages into 4 categories; call establishment, call information phase, call clearing, and miscellaneous. Bit 8 is not currently used. Several other information elements, required for each message type, may be included after octet 4, the exact content of each message being dependent on the type of message. The coding rules are not fixed and additional information elements can be added to satisfy any requirement added in the future.

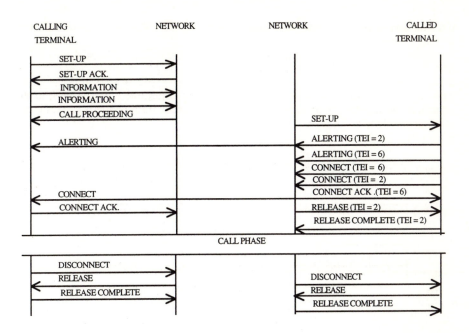

Fig. 5.12 *Basic call establishment and clearing message sequences*

To make a basic ISDN call either overlap sending or *en bloc* can be used. If overlap sending is used then the SETUP message (Fig. 5.12) sent over the 'D'-channel must contain the bearer service request, but called party number and facility requests can be provided in a sequence of INFORMATION elements after the SETUP message. When sufficient information to establish a call has been received by the network, a CALL PROCEEDING will be returned to indicate that the network is attempting to establish a connection.

At the called side, a SETUP message is delivered to all terminals connected to the NT1 via a broadcast message. Each terminal can examine the message to ascertain whether they are compatible with the calling terminal. The compatibility check is done by examining the bearer capability and the lower layer compatibility information element. The application requested (e.g. Group 4 Facsimile) should be contained in the higher layer compatibility information element but this not always the case. In Fig. 5.12, two terminals (TEI 2 and TEI 6) have confirmed that they are compatible and both return ALERTING messages to the network. Each should generate an incoming call indication (e.g. ringing). When the call is answered on either telephone (i.e. handset lifted on a telephone), then that terminal will forward a CONNECT message to the network. The first terminal to forward a CONNECT will receive a CONNECT ACKnowledged message back from the network, instructing the terminal to connect to the correct 'B'-channel. Included in the CONNECT ACK message will be the 'B'-channel's identification. All other terminals which sent CONNECT messages will receive a RELEASE message from the network instructing them to clear and return to the idle state. Following receipt of a CONNECT message from a called terminal, the network will advise the calling terminal that the call has been answered by sending a CONNECT message. Either end can clear at any time by sending a DISConnect message and, on receipt of a RELease from the network, transmission of a RELease COMPlete from the terminal.

5.9   Addressing a specific terminal

The user making a call may decide to make a call to a specific terminal, e.g. when making a call to a user having several telephones connected to an ISDN access, the caller wishes to ring only one particular telephone. Two methods exist in ISDN basic access for addressing a specific terminal. They are Multiple Subscriber Number (MSN), or Direct Dialling In (DDI) , and SUB-addressing (SUB).

5.9.1   MSN

When the MSN supplementary service is provided, the last digit(s) of the network number is sent in the incoming set-up message to all terminals on the user-network interface. Each terminal will have been allocated a different MSN number which is programmed into the call handling process in the terminal. Only the terminal with a matching number will respond to the SETUP message, provided the terminal is compatible (supports MSN) and is in a state to accept the call.

5.9.2   Sub-addressing

In this case, additional addressing information (sub-address) is sent transparently from the calling terminal to the called terminal. The address is not part of the ISDN number used to route the call but is additional to that number. Again, each terminal will be allocated a sub-address which, again, will be programmed into the terminal. Only the terminal with a matching number will respond to the incoming call.

5.10   The current world situation

ISDN is available in virtually all the countries having a major telecommunications network. The main problem in getting a worldwide ISDN has been caused by the differences in implementations by the various network operators, leading to incompatibilities. These incompatibilities are gradually being ironed out with a common approach to standards by the world leaders in ISDN, mainly in the USA, Japan and the major Europe countries. Many countries now have international ISDN interconnection; in early 1994 the UK had connection to over 25 operators. This was expected to increase by an average of almost one a month during 1994/95 (Fig. 5.13). The rate of interconnection internationally will always depend on each country having an internal ISDN network, international switching capabilities and links to other countries (satellite, land lines, undersea cable, etc.). This all requires a lot of money and manpower, as well as expertise, resulting in poorer nations taking some considerable time to join in the international ISDN 'circle'.

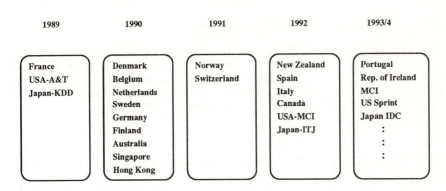

Fig. 5.13 *List of BT connections to international operators (early 1994)*

5.11 The current European situation

The European Commission have recognised the importance of a 'standard' communications service in Europe and that it would be to the benefit of all member countries if all European network providers offered a compatible service making interworking between each country much easier to address. Agreement was reached by the signing of a Memorandum of Understanding (MoU) in 1986 by twenty-three network operators from 17 countries on the implementation of ISDN in Europe. The MoU states that all the countries involved will provide an ISDN service based on a minimum set of requirements by 1993.

The minimum requirements are:

- An international interface based on both the CCITT Signalling System #7 ISUP and TUP+
- A user to network interface compliant with the current European standard for Basic Rate Access
- A user to network interface compliant with the current European standard for Primary Rate Access

A minimum offering of services to be:

- Circuit mode 64 kbit/s unrestricted bearer
- Circuit mode 3.1 kHz audio bearer

A minimum offering of supplementary services to be:

- Calling Line Identification Presentation (CLIP)
- Calling Line Identification Restriction (CLIR)
- Direct Dialling In (DDI)
- Multiple Subscriber Number (MSN)
- Terminal Portability (TP)

To ensure that a common standard exists throughout Europe, the European Telecommunications Standards Institute (ETSI) produces European

Telecommunications Standards (ETS). These are then submitted to the Technical Regulations Applications Committee (TRAC) for adoption as 'CTRs' (Common Technical Regulations). CTRs then become mandatory in each member state. CTRs replace the old NETS, (a change in terminology to replace the rather confusing situation where NET was the abbreviation of the French translation of ETS). In the past each country has had its own network, making international interconnection difficult, with countries having to reach agreement on which signalling system to use. To change the communications network in each country would be an impossible task, but ensuring that standard interfaces exist between user and network, and an international standard exists between member states, ensures that the users can be guaranteed interworking from one country to another.

At the end of 1993, the Euro-ISDN was launched. In some countries, such as Germany and Belgium, the national version can continue to be used to the year 2000 in parallel with an Euro-ISDN offering using a national to Euro-ISDN converter in the network. In the UK, BT had by the start of 1994 over half of Europe's existing Euro-ISDN basic rate connections (nearly 13,000), offering the service to over 80% of the UK business market with connection to the network within six working days. In contrast, the German and French carriers only launched a Euro-ISDN service in late 1993 to comply with the MoU, both countries having had a national version of ISDN since the mid 1980s. Between them, France and Germany have more than 99% of the basic rate connections in Europe based on non-Euro-ISDNs.

5.11.1 Primary rate access

In late 1993, BT launched a Euro-ISDN primary rate service, although the national variant based on DASS2 will continue to be offered to at least the year 2001, or for as long as users have a requirement for it. Mercury in the UK also trialled a Euro-ISDN primary rate service in late 1993.

5.12 The main users of ISDN and what they use it for

A study by BT has shown that the main users of its basic rate ISDN network in 1992 were manufacturing and communications, followed closely by distribution/retail and services, as shown in Fig. 5.14.

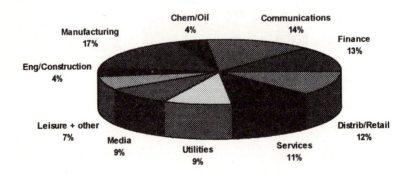

Fig. 5.14 *Industry split of basic rate, September 1992 (BT)*

BT's figures for primary rate users, Fig. 5.15, shows that the finance sector is the largest user of primary rate in the UK, with finance utilities and distribution/retail making up 50% of all users.

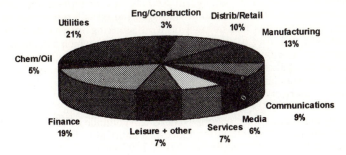

Fig. 5.15 *The industry split of primary rate, September 1992 (BT)*

5.12.1 *International use of ISDN*

In 1992 BT found that internationally the main use of ISDN was for Video Conferencing or video telephony (50%), with KiloStream Back-up the in second place with 10% (see Fig. 5.16). The reason for the large amount of usage for video traffic is attributed to many factors, one being an international standard for video which is easily transported to the ISDN and, a second, that many companies saw the advantage and cost saving of using videoconferencing. Cutting back on time wasted in travelling and a cost saving of more than a sixth on current charges when the cost of three business class tickets to the USA is compared to the cost of six basic rate ISDN channels carrying a video conference.

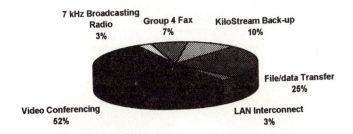

Fig. 5.16 *BT's international user base, December 1992*

Fig. 5.17 shows the main usage of ISDN products attached to BT's network in the years 1991-93 . In the third figure (1993) Group 4 Fax has not been shown as a main application, but indications show that it is perhaps a second or third application. It could be assumed that the use of ISDN for KiloStream back-up will decrease as the users requiring this service will reach saturation.

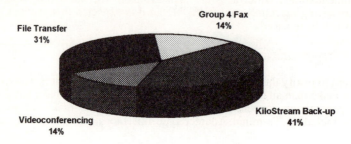

August 1991

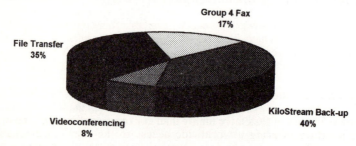

April 1992

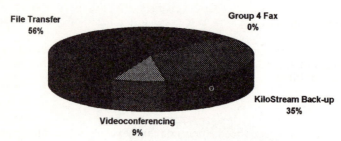

April 1993

Fig. 5.17 *ISDN basic rate application usage (BT): Source NBES Survey April 93*

5.13 Available types of Customer Premise Equipment (CPE)

5.13.1 Products available

There are many products now becoming available for use on the ISDN including software applications that run on PCs. Together these provide the many ISDN applications we see used today. The lists below show some (by no means all) of the products, and some applications available now.

5.13.2 Terminal Adapters (TA)

These provide the rate and protocol adaptation of existing products based on V- and X-series Recommendations. Interfaces are also available for interfacing analogue telephones, answering machines, etc. to ISDN, but it is expected that eventually all these products will be replaced.

5.13.3 ISDN telephones

It has been hard to justify the need for a basic ISDN telephone since it is many times more costly to make than an analogue telephone and ISDN will not make voice calls cheaper by speeding up speech! Manufacturers have had to make the product acceptable by including Terminal Adapter type interfaces and include attractive features.

5.13.4 The ISDN Private Branch Exchange (isPBX)

Primary rate and proprietary interfaces have been available for some time on isPBXs. Basic rate interfaces for PBXs have been talked about for many years but it has taken until 1993 for approved versions to start to appear. The isPBX offers many possibilities over the traditional PBX such as packet handling or LAN bridging.

5.13.5 Personal computer cards

ISDN will promote the use of PCs more highly than any other application. The offer of a low cost interface offering a worldwide access to the user is remarkable. Worldwide access offers the user a doorway to a vast database of knowledge, undreamed of prior to ISDN.

5.13.6 Multimedia terminals

Usually based on a desktop computer, users can talk, watch and transfer data, usually over two channels of ISDN. Many see the multimedia terminal as the best cost-effective terminal for ISDN.

5.13.7 Group 4 fax

For users who have a continuous need to transmit fax messages, the savings gained by moving to an ISDN fax must be carefully balanced against the cost of the equipment and ISDN line and the amount of usage. For the press the greater resolution and speed of transmission with improved quality of picture makes this very attractive.

5.13.8 Video telephones and videoconferencing

The new products being launched now offer tremendous quality and cost savings when compared to previous products. If anything will justify the costs involved with ISDN these products will. The world has become addicted to screen-watching and it would be difficult to visualise a future in communications without video playing a major part. Most video terminals offer working over two 'B'-channels using video and speech compression, PC card versions also include data transfer. Videoconferencing usually takes six basic rate channels for improved picture quality. Products vary from simple video phones through PC-based multimedia terminals to full videoconferencing suites.

5.13.9 Bridges and routers

Used to interconnect or access LANs, many businesses interconnect their offices using routers, which can be a combination of both basic and primary rate.

5.13.10 Premise control units

Typically these provide a point of sale function and alarm system connection; they can be used for remote working from home or remote home control.

5.13.11 Multiplexors and inverse multiplexors

These come in many shapes and sizes; they offer combinations of N x 64 kbit/s, either multiplexing up or down. Uses are for videoconferencing, modem replacements for high speed file transfer, and LAN/WAN bridging.

5.14   Typical uses for ISDN

5.14.1 Telecommuting

Extending the office LAN out to a home has the advantage of saving travel and office space. ISDN allows electronic mail, file transfer and videoconferencing, as well as the traditional fax and audio access. The additional benefits of home banking, home shopping, home health care, information services access, remote gas, water and electricity meter reading, fire and burglar alarms, all fed over one copper pair, must surely change lifestyle in the future.

5.14.2 LAN interconnect

The standard method of linking LANs by private circuits or packet networks, nationally, or even internationally, is expensive. ISDN makes it cheaper and easier. Bridges and routers can interconnect Ethernets and other LANs, opening up access to large quantities of information.

5.14.3 Electronic Point Of Sale (EPOS)

Today's cash registers are virtually PCs with a drawer for keeping money in. An ISDN PC card can provide a means of downloading any information taken on the till to a central mainframe, offering stock assessment, price adjustment and card verification in seconds. Additional facilities such as burglar, fire and temperature alarms, meter readings and video verification are some of the more obvious.

5.14.4 Video

The ability to receive video transmission over the local copper loop is extremely attractive; applications for video seem endless. How long will it be before a simple audio telephone used today will be a thing of the past, like 405 line, black and white television?

5.15   Future

ISDN has not had an easy life so far. Lack of standards and the resulting lack of terminals, coupled with the high cost of the provision of an ISDN network, have all contributed to ISDN's slow take-off. Even today there are still threats; claims are being made of modems capable of a throughput of over 100 kbit/s using the CCITT V.FAST specification and V.24bis, i.e. data compression techniques. However, ISDN's basic rate channel throughput of 64 kbit/s consists of raw data and, using the same compression techniques, this could be extended to over 200 kbit/s.

An N x 64 kbit/s channel aggregation service offering users bandwidth on demand, or multi-type interfaces through one access would seem attractive to both user and provider. The network operator's vision to integrate all services delivered over digital networks into one ISDN access will require not just the provision of basic rate or primary rate channels, but offering to the user the option of as many channels as required, i.e. a capacity from 64 kbit/s to 2048 kbit/s in increments of 64 kbit/s.

The demand for increases in data rates is sure to continue. Within the next five years broadband ISDN should be available, offering data rates in the order of 100 Mbit/s. Even this will not be enough; if developments such as 3D-TV are to be routed through the network then maybe data rates greater than 200 Mbit/s will be just as common as 64 kbit/s in the next century.

ISDN today offers a high performance, very flexible dial-up connection. It offers almost instant visual and audio connection to most major countries in the world. ISDN does not claim to be the perfect answer to all communications problems, but it does provide a very good platform for solving most of them.

Chapter 6

Broadband ISDN

Ian Gallagher

6.1 B-ISDN

Broadband ISDN was conceived as a network able to support the full range of imaginable services from low bit-rate telemetry to high definition image and television transmission including voice, video, data and multimedia. A key requirement was a high degree of flexibility on a number of timescales: long term to cope with uncertainties in services and their demands on the network, short term to cope with mobile customers and terminals, and instantaneous as customers switched connections and services. For its introduction, B-ISDN would rely on the widespread availability of high bandwidth optical fibre transmission, not only as trunks between switch nodes but also in the access to customers' business premises and homes. B-ISDN would also require the development of new signalling and control systems to support the flexibility and allow the customer to request changes to his services and connections.

As a concept and goal for the future, B-ISDN seems to be well accepted and much activity is being directed toward identifying and resolving the issues and developing the required standards. However, it is much less clear when and in what form a commercial case could be made for the required investment as the economic, market, technological and regulatory pressures have a complex and changing relationship with each other. Probably it will be into the next century before B-ISDN is generally available.

In the shorter term, over the next three to five years perhaps, the situation is becoming a little clearer. There are signs of an early business requirement for higher bandwidth services, especially for data. The current view is that these could be met using the switching and multiplexing technique, selected as the basis for the future B-ISDN, known as Asynchronous Transfer Mode (ATM). Thus ATM, having been conceived as a solution for B-ISDN, has now taken on a life of its own.

6.2 ATM basics

ATM is a technique based on the use of short, fixed length packets (known as cells) to support all voice, video and data services. As shown in Fig. 6.1, the cells are 53 bytes long, including a 5 byte header and a 48 byte information field. The header carries sufficient information to route the cell across the network so that cells from different terminals can be multiplexed together on the same bearer. On the bearer, the cell stream is continuous and the cells are filled asynchronously as the terminals have data to send. At the switch, the cell header is used to address a look-up table to determine the switch outlet required. The contents of the look-up table are pre-set by the management and control system. Since the switching is done in hardware, very high throughputs can be achieved; of the order of 10s or even 100s of gigabits/second are possible with today's technology.

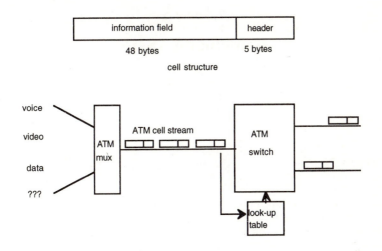

Fig. 6.1 *ATM structure*

The information field length of 48 bytes was chosen in 1988 as a compromise between the American desire for 64 bytes to better suit data and the European desire for 32 bytes to suit voice. The issue for voice is one of delay in building up the information field. At 64 kbit/s it takes 6ms to fill 48 bytes. This delay may or may not be acceptable without some form of echo control, depending on the loss and round trip delay of the connection. The issue for data is one of efficiency and the ability to carry the complete header of a higher layer data packet within one cell information field.

The cell header consists of a number of fields as shown below:

Header Error Check (HEC)	1 byte
Virtual Circuit Identity (VCI)	2 bytes
Virtual Path Identity (VPI)	1 byte (12 bits at Network Node Interface)
Generic Flow Control (GFC)	4 bits (not at Network Node Interface)
Cell Loss Priority (CLP)	1 bit
Path Type Field (PT)	3 bits

The HEC field provides error checking on the header only. Any checking on the contents of the Information Field is provided separately.

The VCI and VPI fields identify individual connections within a transmission pipe as illustrated in Fig. 6.2. A virtual path (VP) consists of a bundle of virtual connections (VC) with the same endpoints. For example, a VP could be set up from an ATM switch in Ipswich to one in Birmingham with individual connections between customers in each town being allocated VCs. A feature of VPs is that they can be elastic-sided with their bandwidth varying as the VC bandwidths vary. By setting different boundaries on VPs, bandwidth can be traded or reserved as required for different services. They can be used to limit the effect of one service on another should they encroach on each others' bandwidth. By switching different VPs to different queues at the inlet to a transmission system it would be possible to allocate different delay priorities.

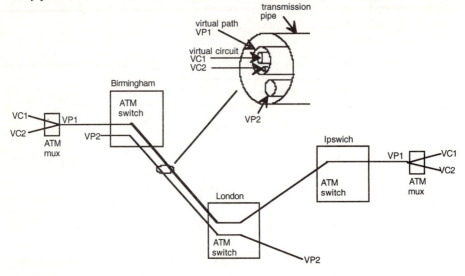

Fig. 6.2 *Transmission pipe structure*

The use of the GFC field is still not fully defined. The intention is to use it to control the access of terminals to the network across the user network interface (UNI). It is therefore not needed on the network side of the switch at the network node interface (NNI).

The cell loss priority bit is intended to indicate those cells which should be discarded in preference to others at points of congestion. However, the exact circumstances of its use are still not defined.

The path type field is used to identify special ATM cells which have the same VPI/VCI values as "normal" cells and hence follow the same route and encounter the same performance. They are therefore used for maintenance purposes in checking the integrity and performance of a particular connection or providing congestion notification.

It is clear that the different services to be supported on an ATM network could have widely different requirements. Real-time services such as voice are very delay-sensitive but tolerant of loss. Data on the other hand is more tolerant of delay but very sensitive to loss. Data requires more sophisticated error protection which would be redundant for voice. Hence a range of methods for adapting services into ATM cells has been developed and, in layer terminology, they form the adaptation layer sitting above the ATM layer.

Below the ATM layer lies the physical layer and standards have been developed for packing cells into most of the well known transmission systems including the emerging SDH systems at 155 and 622 Mbit/s as well as the existing PDH systems at 34 and 140 Mbit/s. The structure of these layers is illustrated in Fig. 6.3.

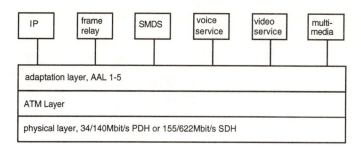

Fig. 6.3 *ATM layer structure*

6.3 Services and interfaces

To be introduced into the network and offer a sensible evolution path, ATM must be able to offer support for a wide range of existing services and interfaces. These include the alphabet soup of SMDS, CBDS, DQDB, MAN, FR, CBR, VBR, PVC, SVC and others. Taking these in some sort of order:

MAN   Metropolitan Area Network. Originally a network that provided services over the area of a city. A MAN standard was developed by IEEE 802.6 (called DQDB) with the intention of supporting a high speed connectionless data service. Fig. 6.4 illustrates the relationship between the various acronyms.

DQDB   Distributed Queue Dual Bus. A protocol for terminals accessing a shared dual bus. It is based on ATM size cells with a very similar header such that conversion is relatively straightforward.

SMDS   Switched Multimegabit Data Service. Bellcore definition of a connectionless data service originally developed for use with DQDB-based MANs. However, it can also be supported on ATM-based networks. A European version of SMDS is CBDS, Connectionless Broadband Data Service.

Frame Relay   A connection-oriented interface definition which grew out of standards to support packet data over ISDN. It was first implemented commercially as an interface standard by private circuit (e.g. T1) mux suppliers who set up the Frame Relay Forum in the USA to develop the specification.

CBR   Constant Bit Rate. This term refers to emulating traditional circuit-based services such as those at 64kbit/s or 2Mbit/s over an ATM network. One of the features of such a service is the need to offer some sort of guarantee of bandwidth.

VBR   Variable Bit Rate. This term usually refers to real-time services such as voice or video where cells are only generated when necessary. That means, for instance, not sending cells when there is silence in a voice connection or when nothing is moving on a video connection.

PVC   Permanent Virtual Circuit. This is the equivalent of a traditional private circuit. It is set up by the operator at the request of the customer to provide a permanent connection for the customer to use as he wishes. The major difference to a traditional circuit is that when not in use the bandwidth could be made available for other services and connections. This type of connection is available on early ATM switches.

SVC   Switched Virtual Circuit. This is the equivalent of the traditional dial-up connection. It is set up by the terminal signalling a request to the call control system. The standards for signalling in B-ISDN are not yet fully defined although an initial release has been agreed and is becoming available.

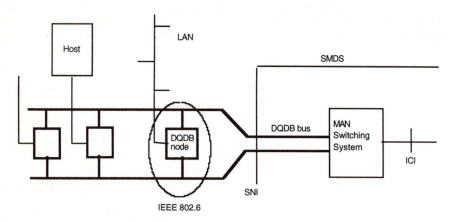

Fig. 6.4 *IEEE 802.6 acronym relationships*

As mentioned above, a range of adaptation layer types has been developed to support this range of services. There are now five types defined, AAL1 - 5, of which AAL1- 4 were the original definitions and AAL5 is the newcomer.

AAL1   For CBR services. Intended to support circuit emulation, it has a one byte overhead containing a sequence number to detect missing cells.

AAL2   For VBR services. Intended to support variable bit rate video or voice, it is not fully defined yet.

AAL3   For connection-oriented data. Originally this was intended to support services such as frame relay and also signalling. After further study it was agreed to combine AAL3 with AAL4 because of their similarity and so AAL3/4 was defined. However, a more efficient AAL has now been developed particularly for data services and referred to as AAL5.

AAL4   For connectionless services. Originally this was intended to support services such as SMDS or CBDS. As mentioned above, it was superseded by AAL3/4 which currently is the accepted AAL for SMDS.

AAL5   Efficient AAL. AAL5 has become the accepted AAL for frame relay and signalling. It is also under study as the AAL for protocols such as the Internet Protocol (IP) and LAN networks such as Novell's IPX. Whether it becomes the standard for connectionless services such as SMDS remains to be seen.

6.4   ATM networks

As shown in Fig. 6.5, a basic ATM network can consist of ATM switches and multiplexors interconnected by high speed trunks and monitored and controlled by a network management system. Initially the switches will be simple cross-connects set up by the management system to provide permanent virtual circuits between endpoints of the network. A feature of ATM networks is the buffering required at the inlet to each transmission link. It is the control of traffic through these buffers that has led to many of the unresolved issues in ATM standards. Based on equipment becoming available from the large public network suppliers, the switches tend to have of the order of 64x64 155 Mbit/s ports and a throughput of 10-100Gbit/s. Because there are not many SDH systems installed, and to limit the expense on bandwidth, it is possible to adapt the ports to 34 or 140Mbit/s. Multiplexors to gather together traffic from different sources are also available. They may contain adaptation functions for a variety of services and interfaces and may also provide some bandwidth gain by statistical multiplexing.

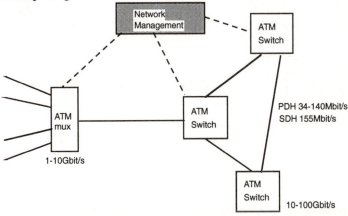

Fig. 6.5 *ATM network*

Fig. 6.6 illustrates the support of frame relay across an ATM network. It shows a LAN connected through a router to a frame relay UNI on a multiplexor where the frames are adapted, probably using AAL5, for transport across the ATM network. Part of the adaptation function is to translate the frame relay DLCI into an ATM VPI/VCI to

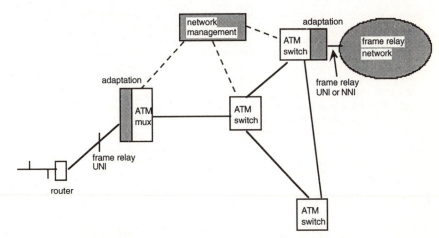

Fig. 6.6 *Support of frame relay on ATM network*

route the cells to the correct piece of adaptation equipment at the distant end. This translation must be agreed beforehand with the customer and set up by the management system. At the distant end the the original frame is reassembled for delivery to a terminal or a further frame relay network.

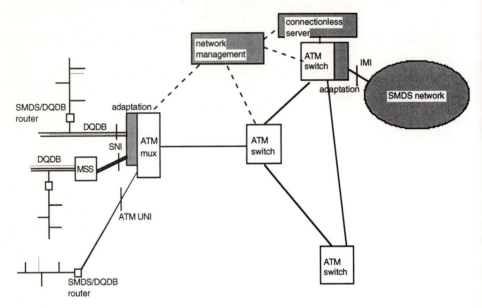

Fig. 6.7 *Support of connectionless data services*

Fig. 6.7 illustrates the support of connectionless data services such as SMDS. It shows LANs connected by a number of methods, using MANs as the local access mechanism into an ATM network. Briefly the three methods shown are:

* A LAN connected via an SMDS router onto a DQDB bus providing a Service Network Interface (SNI) to the adaptation equipment in the ATM multiplexor.
* A LAN again connected to a DQDB bus but which is then combined with traffic from other DQDBs in a MAN Switching System (MSS) before connection to the adaptation equipment. The interface between the two could either be an SNI or an Inter Carrier Interface (ICI) depending on whether it represented a boundary with a customer or another network.
* A LAN connected directly to the ATM network by means of an SMDS router with an ATM output instead of DQDB. The adaptation function is carried out in the router rather than at the multiplexor.

SMDS is a connectionless service using 60 bit E164 addresses to identify terminals on a global basis. Obviously to make full use of this ubiquitous address it would not be possible to preallocate enough VPI/VCI values. Therefore the adptation function segments the higher layer packet into ATM cells and uses a common VPI/VCI which routes all cells to a connectionless server. The payload of the first cell contains the source and destination E164 addresses and the server translates this to a VPI/VCI for the next stage of the journey either to a terminal, an SMDS network or another connectionless server. The following cells of the packet have a message identity in their payload to associate them together and have the same VPI/VCI added. In addition to the switches and multiplexors, the network management now has the address tables in the server to manage.

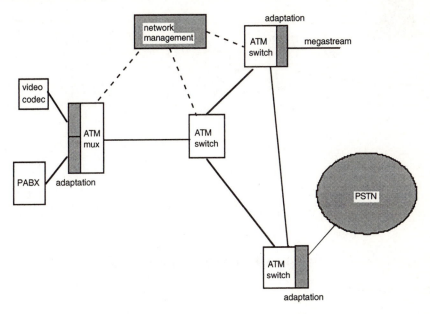

Fig. 6.8 *Support of constant bit rate services*

Fig. 6.8 illustrates the support of constant bit rate services. It shows CBR sources such as a PABX with 2Mbit/s trunks and a video codec via an ATM network. The problem for the adaptation function is to ensure that the timing is regenerated correctly at the distant end. If the clock frequencies are different then bits may be lost or inserted in a manner whch affects the service being supported. Methods based on using an adaptive clock and/or a system referred to as Synchronous Residual Time Stamping (SRTS) are under investigation. The problem for the ATM network is to ensure that enough bandwidth is guaranteed and available for the services. This will be discussed further under the following section on ATM issues.

Fig. 6.9 illustrates the support of ATM CPE interconnection. It shows LANs and workstations being interconnected using ATM Hubs as premises equipment, providing local interconnect as well as access to the wide area. ATM Hubs are a recent phenomenon, emerging rapidly as a result of work by the ATM Forum, originally a group of suppliers who gathered together to define interfaces to ATM CPE. The work of the Forum is discussed further under the section on the international position.

From the above, it is clear that there are several ways of providing interconnect between LANs as long as they are connected to the same service. Interworking between LANs with frame relay routers and those with SMDS routers has not yet been fully addressed, however. In theory it is possible to carry frame relay over SMDS and vice versa, as well as over ATM. There is obviously a limit to this type of nesting if any sensible performance and efficiency is to be achieved.

In addition to the services mentioned above, which perhaps are more associated with the public network operators, the internet community is well on the way to generating standards for operating IP over ATM and addressing the problems of routing data over very large networks.

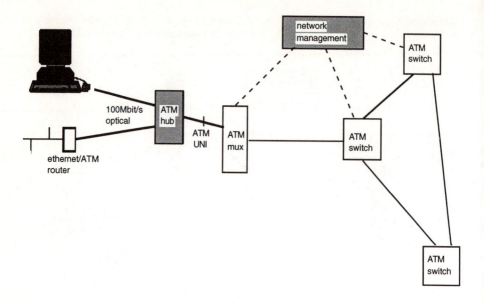

Fig. 6.9 *ATM CPE interconnection*

6.5   Signalling

CCITT and ETSI are developing B-ISDN signalling through successive phases, called
Capability Sets (CS). CS 1 allows the support of existing telecommunication services,
such as voice and other constant bit rate services, using ATM. CS2 and 3 will
introduce new and enhances services. A summary of the features of each is given in
Table 6.1.

The similarity between N-ISDN and CS 1 services has resulted in the current
signalling systems, ISUP for the network and Q931 for the access, being modified to
provide B-ISDN CS 1 features. The equivalent broadband signalling standards are
therefore B-ISUP and Q2931.

The ATM Forum released their access signalling specification in mid-1993, based
on Q2931 but with some differences. These differences have now been overcome and
the ATM Forum specification and ITU Q2931 CS 1 are now aligned.

The discussion so far has been concerned with the support of individual services.
However, a number of significant issues arise when combining them together on one
ATM platform to form a multi-service network and the starting point for B-ISDN.

6.6   ATM issues

Most of the difficult technical issues surrounding the implementation of ATM
networks concern traffic control. Fig. 6.10 illustrates the significant set. Starting at the
edge of the network they are: the use of generic flow control, cell delay variation,
policing/shaping, connection admittance, congestion notification and the effect of
connectionless traffic. They are not independent, so each is affected by the others. The
aim in solving these issues is to be able to offer services to the customer with some
sort of guaranteed defined quality, capable of being verified.

CS1	CS2	CS3
Available now	Available 1995	Available post 1995
Peak bit rate connections	Variable bit rate connections	Multimedia and distributive services
	Quality of service indication by the user	Quality of service negotiation
Point-point connections (uni- or bi-directional) (symmetric or asymmetric)		Multipoint connections
Signal connection established on demand	Multi-connection, delayed establishment	
	Use of cell loss priority	
Peak rate allocation only	Negotiation and renegotiation of bandwidth Bandwidth allocation based on traffic characteristics	
Point-point signalling access	Meta-signalling for point to multipoint signalling access	
Limited supplementary services	Supplementary services	

Table 6.1 *B-ISDN capability sets*

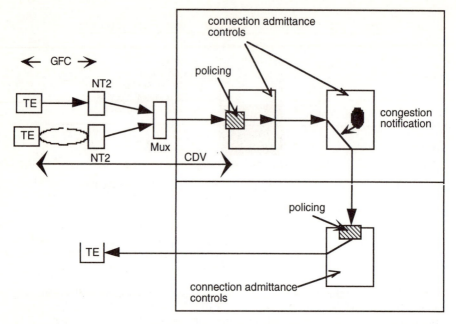

Fig. 6.10 *ATM traffic control features*

Generic Flow Control (GFC) concerns the use of four bits in the ATM header to control terminal access. The principal way in which these four bits will be used has been decided by the ITU and a new version of I150 is expected to be approved in the near future.

Cell delay variation (CDV) is an effect caused by the different delays experienced by cells as they pass through the various buffers within multiplexors and switches. Thus, for a constant bit-rate (CBR) service generating cells at constant intervals, the cells may become bunched or separated at an early stage of the network. This makes policing the cell rate more difficult. It also means there is a higher chance of buffer overflow if, for instance, a number of CBR services have their cells bunched at the same time. There is also evidence to suggest that high bit-rate CBR services may experience high levels of CDV when mixed with several low bit-rate connections.

Policing is required where the traffic on a customer's access link is combined with others at a multiplexor or switch. The purpose is to inspect the rate of cell arrivals and take action to prevent any excess, over the rate agreed with the customer, affecting other traffic by pre-empting its bandwidth. Policing is carried out by means of a leaky bucket as shown inFig. 6.11 The depth of the bucket is set to allow for clusters of cells arriving together due to the effect of VBR traffic and CDV. If the bucket becomes full further cells may be discarded immediately or tagged, using the CLP bit, for possible future discard. Note that the cells themselves do not pass through the bucket so there is no extra delay. Selecting the depth of the leaky bucket for CBR services and VBR services policed at peak bit-rate seems possible even allowing for CDV. However, for VBR services where statistical multiplexing is required, the problem is not solved. A number of suggestions are under consideration, including relying on terminals with known characteristics or explicitly defining the traffic as a maximum sustained rate. Neither is very satisfactory and both rely on the customer providing accurate information in his connection request. For the present, it appears that all services will be policed on peak bit-rate.

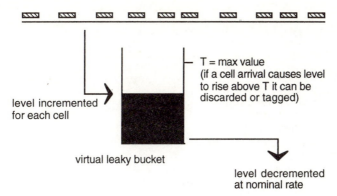

Fig. 6.11 *Leaky bucket*

If policing is based on peak rate, then call requests need only be checked by calculating the spare bandwidth and comparing that against the bandwidth requested. Call acceptance algorithms for the case where statistical multiplexing is required are still under study.

The CLP bit is currently defined to be set only by the network to indicate that the cell is eligible for discard. However, the debate has recently been re-opened particularly in the ATM forum, where some terminal suppliers wish to use CLP to indicate high and low priority cells.

In networks, such as those using frame relay, congestion can be notified to the sending terminals by means of special bits in the header called forward or backward explicit congestion notification (F or BECN). A similar indication has been included in the ATM header as one of the path type field values. However, there are a number of issues to resolve regarding its use. It is not effective when congestion is of short duration or if some terminals are unable to react. The terminal causing the congestion is difficult to identify and may not be the one which is suppressed.

Finally, there are several traffic control problems associated with connectionless traffic. The traffic on a link can fluctuate rapidly and unpredictably, making planning and dimensioning the network difficult. Connectionless works well in a LAN-like environment where utilisation is low and so, in a broadband public network, application bit-rates need to be low compared to the link capacity, otherwise a few customers sending large bursts will cause congestion. One method of improving the quality of service may be to segregate connectionless traffic onto its own links or VPs.

The traffic control issues just described present significant problems. However, it is not all bad. To summarise:

If connections are all CBR then ATM traffic controls are simple and covered by I371.
If VBR connections are given peak rate allocation then I371 also applies.
Many of the issues surrounding statistical multiplexing are not unique to ATM i.e.
    how to assure the QOS
    how to police connections
    how to provide congestion notification
    how to provide shared access to a UNI.

All these issues will be the subject of an enhanced version of I371 scheduled for completion by 1994.

It is clear that just enough is understood about the traffic control issues to make a start in implementing an ATM network. Perhaps, too, there are enough potential customers with the applications to make a start worthwhile.

6.7   Applications

The list of applications for broadband normally contains items such as

> LAN interconnect
> High speed file transfer, especially image and back-up files
> Video
> Co-operative working
> Service integration, voice, video and data
> Multimedia

to which could be added for the future, signalling for Intelligent Network applications and mobile applications.

As can be seen by the list of applications, ATM will tend to be introduced to serve large business users initially rather than the residential market. Therefore the deployment of equipment is likely to consist of only a small number of switches located in major business centres connected by high capacity links based on the SDH network. It is not seen as a replacement for the existing PSTN and ISDN, at least in the short term. Having said that, there is a growing view that ATM has a part to play in the provision of video-on-demand and other broadband services to residential customers.

There is no single killer application identified as being the one that must use ATM. All the applications could be implemented by means of various combinations of other solutions. However, ATM is seen as the most promising way of realising one very flexible solution for a whole range of applications and services.

There are perhaps three sets of drivers for implementing some form of ATM network.

6.7.1 To obtain a more efficient network infrastructure

It can provide one network infrastructure to support a range of services, in particular, high speed data including SMDS and frame relay. The packet switching nature of the network means it is possible to support bursts of data between different sets of destinations without having to purchase or rent a physical mesh of high speed links.

6.7.2 To interconnect ATM CPE

Over the last year there has been great interest, expressed through the ATM Forum, in ATM LANs. They offer higher performance, greater flexibility and lower cost than current LANs. It appears that organisations may instal ATM LANs on their own merits and will soon want to interconnect over the wide area, thus encouraging the development of public ATM services for permanent, and later switched, virtual connections.

6.7.3 To be better prepared for future opportunities

There are many possible opportunities for the future. In the video services area an ATM network allows the trade-off between picture quality, coding, bandwidth and switching to be explored without the need to change the network. The self-steering nature of a packet network, and its multicasting ability, lend themselves to some of the problems of conferencing, mobility and signalling to a range of servers. There is no clear view of what services customers will want to pay for in the future nor what those services might require of the network. ATM, more than other techniques, has the flexibility to cope with that uncertainty.

Following on from what could happen, the next section is concerned with what is happening around the world now.

6.8 The international position

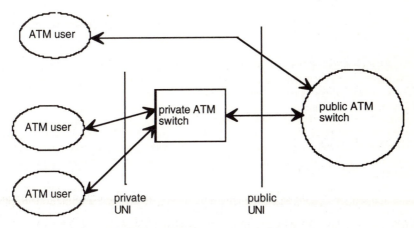

Fig. 6.12 *ATM Forum areas of interest*

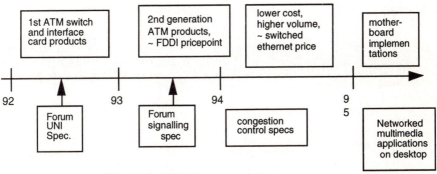

Fig. 6.13 *ATM Forum activity programme*

Since 1986 most of the work in CCITT and ETSI on B-ISDN has centred on standardising the public network and, in particular, the User Network Interface (UNI). For customer premises equipment, the major force which has emerged is the ATM Forum set up by the industry in the USA in 1991 to define interfaces to ATM CPE. Today it has over 300 members from all sectors of the industry and has a very commercial focus. Fig. 6.12 shows the ATM Forum's areas of interest. The public UNI is the same as the CCITT version. The Forum produced issue 1 of the private

UNI specification in June 1992 and released issue 2, including the UNI signalling specification, in mid-1993. A number of companies in the bridge/router and T1 mux. market are now selling ATM CPE, based on the Forum specifications. The attraction to customers is the higher performance, greater flexibility and lower cost than other solutions such as FDDI. A time line for possible deployment suggested by the Forum is shown in Fig. 6.13.

All of the traditional large public network switch suppliers have developed ATM switch hardware and are now developing the associated management and control systems. Many of them will have equipment available for trials by the end of this year and most operators in the USA and Europe are now planning such trials. In Europe a group of 18 Public Network Operators have signed up to implement a Pan-European ATM network. It is based on each operator procuring at least one switch interconnected to the rest of Europe by 34 or 140Mbit/s links. The objectives are:

to test the capability of the technology to support a range of services including SMDS and frame relay

to test interworking between different operators' and suppliers' equipment

to investigate traffic control issues

to learn about the management and control required

to demonstrate applications

to allow pilot users to assess the network.

The equipment is scheduled for installation in the UK in December 1993 and ready for service in July 1994. One of the pilot users will be the UK universities broadband network, Superjanet, to which several universities are attaching their own ATM switches.

In the USA, several 'gigabit' network projects have been established under government agency funding to experiment with high speed applications and technology. In addition, several US operators have announced the availability of ATM services.

Largely because of commercial organisations and commercial pressures, particularly in the USA, ATM is beginning to move. Unlike ISDN, terminal equipment and applications are being developed according to market forces. However, how fast and how far the world will move on towards a full blown B-ISDN remains to be seen.

Chapter 7

Basic LAN structures

A. Gill Waters

7.1. Introduction

The purpose of a Local Area Network (LAN) is to provide interconnection between a variety of computing systems within a building or small site, typically within an area of up to 5 km diameter. This interconnection helps users to communicate effectively (by sending messages or formal documents) and to share resources (by sharing printers, file storage, databases etc.). LANs are now commonplace in many organisations as cheaper, smaller and more powerful systems find their way onto an increasing number of desks.

Most LANs are managed by a single organisation and so, theoretically, the way in which systems connected to a single LAN communicate can be unique to that LAN. This fact led to a number of parallel developments when LANs were first developed and the trend has continued. Early work in the mid 1970s included the development of the Ethernet [1] (a bus-based contention LAN) and the Cambridge ring [2] (a ring topology with access through circulating fixed size slots). Standardisation came later as users began to demand conformance to standard LAN interfaces for computing systems from different suppliers. There were several candidates to choose from and, rather than select a single LAN technology, a number of alternative technologies have been standardised. Most computer suppliers offer interface boards for one or more of the standard LAN technologies and chip sets are available to assist in their construction. There are also a number of proprietary LAN networking architectures, mainly for networks of personal computers.

This chapter is constructed as follows. The remainder of this section presents an overall introduction to the characteristics of LANs and to the principles under which they work. The second section discusses standardisation of LANs. This is followed by a description of the three principal LAN standards: CSMA/CD, the token passing bus and the token ring. This chapter concentrates on each of the three LAN technologies in turn. The following chapter on LAN protocols concentrates on higher layer protocols which are independent of the underlying technology.

7.2 LAN characteristics

The various standard and proprietary LANs all have similar characteristics. LANs are restricted to a small area, such as a building, an office or the site of a small organisation. This results in very low transmission delays and higher speeds and lower bit error rates are more easily attainable than for long-distance links. It is therefore possible for computer systems (often called *stations*) to share access to a common medium in an equitable manner, without suffering long delays in gaining access to the medium.

Shared access leads naturally to simple LAN topologies such as a ring, a tree or a bus which in turn reduces cabling costs when a LAN is installed. Example LAN topologies are shown in Fig. 7.1. Some earlier versions also used a star configuration; this tends not to be the case today, though sub-groups of users may be connected to a single access point (Fig. 7.2a) and a star configuration is sometimes used for easy rerouting and isolation in a concentrator box as shown in Fig. 7.2b. Typical LAN speeds are of the order of 10 Mbit/s, using a variety of media such as twisted pairs or specially shielded coaxial cable. Higher speed LANs such as the Fibre Distributed Data Interface (FDDI) at 100 Mbit/s use optical fibre. FDDI is connected in the form of a ring; other fibre-based LANs which require that all stations hear all transmissions use a star topology with a passive star coupler. A number of wireless LANs have also recently been introduced based, for example, on infrared communication.

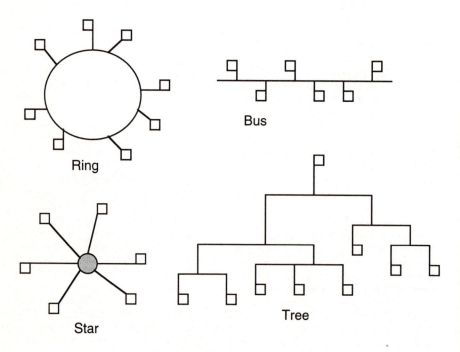

Fig. 7.1 *Typical LAN topologies*

The low transmission error rate and large number of small systems connected to a LAN dictate that LAN access protocols should be reasonably simple and cheap to implement. Error recovery is considered a matter for higher layer protocols, a trend which is now being extended to the wide area, with higher speed packet networks techniques such as frame relay and the Asynchronous Transfer Mode, covered in a later chapter.

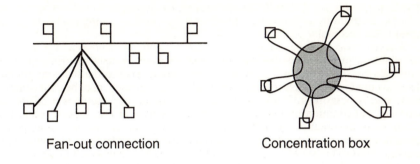

Fan-out connection Concentration box

Fig. 7.2 *Star configurations for economy and reliability*

In summary, LAN characteristics and typical values are:

- low area of coverage ≤ 5 km diameter
- low delay
- low bit error rate $\leq 10^{-9}$
- high speed   10 Mbit/s
- simple topology   e.g. bus, ring
- common medium shared by all stations

7.3 Medium access techniques

Most LANs work by allowing all stations to compete fairly for access to a common transmission medium. This can be done in a number of ways - by contention, by offering access to the network in turn to each station attached to the network or by preallocating bandwidth to some of the stations. Contention techniques can be very efficient if the network is lightly loaded as stations will quickly gain access to the network. However, when working near the capacity of the network the inherent possibility of collisions and therefore of retransmissions increases and access time becomes increasingly unpredictable. Techniques for offering capacity to each station in turn include passing round a token, possession of which gives the right to transmit a frame. Alternatively, as in the slotted ring, fixed size slots are circulated and these can be filled by a station which finds a slot empty; the information is copied by the destination station, but continues round the ring and the slot is marked empty by the sending station thus preserving the original access order to the ring. Allocating fixed capacity is not efficient, since most LAN traffic from data sources is bursty.

7.3.1 CSMA/CD

The contention technique commonly used on LANs has evolved from simpler protocols. The simplest contention technique is always to transmit a frame when data is ready to send. An early packet radio system at the University of Hawaii, called the Aloha system, was based on this principle [3]. A single high speed radio channel was used for incoming frames from many slave stations to a master site. Obviously, collisions could occur. The master would send acknowledgements on a different channel and slave stations could assume that frames which were not acknowledged had collided. They then waited a random time and tried again. The maximum efficiency of such a technique is about 18% of the channel capacity. The original

Aloha system was based on variable size frames. By fixing the size of the frames, it is possible to double the efficiency of the scheme by dividing the time available on the channel into slots and insisting that each frame is transmitted within a slot, a technique called slotted Aloha.

LANs using contention can improve on this system because of their low delay. First, a station can check that no other station is transmitting before launching its own frame, a technique known as *Carrier Sensing*. Secondly, once the frame transmission has started, the station can check the channel directly for collisions - *Collision Detection* - and abort a collided transmission at the earliest possible time after transmission has begun, rather than continuing to transmit a complete frame. Having suffered a collision, stations wait a random amount of time before attempting a retransmission. This is the basis of the Carrier Sense Multiple Access with Collision Detection (CSMA/CD) technique.

7.3.2   Token passing

The token passing concept can be applied to both the ring and bus topologies. A unique token is passed to each station in turn and the station in possession of the token has the right to transmit. Depending on the actual network, it may transmit a single frame, or it may hold the token (and continue transmitting) for a specified amount of time. For token passing rings the token is simply passed to the next station in the physical ring. For token passing buses, a logical ring must be formed and stations must in general be aware of their position in the ring. In both cases the procedures involved in using and passing on the token are very simple, but slightly more complex maintenance procedures are needed to ensure that the token passing mechanism does not run into problems, such as loss of the token.

7.4   LAN standards

As previously mentioned, there have been, and indeed still are, diverse solutions to the provision of LANs. A totally independent solution is not acceptable where users wish to mix equipment from different vendors on the same LAN or where there is a need for off-site interconnection. Standards activity for LANs has tended to lag behind that for WANs, where mixed equipment was more common.

Standards for LANs have been developed by the IEEE project 802 committee. Since it was clear from the outset that it would not be possible to recommend a single standard for LANs, the first job was to define a framework within which the various LAN technologies could be contained and which would also provide a unified approach just above the medium access layer thus enabling higher layers of the OSI model to use LANs in a sensible manner.

With these objectives in mind, the committee produced a series of standards which are related as shown in Fig. 7.3. The standards are labelled IEEE 802.x; the first, IEEE 802.1 defines the framework for the LAN standards and its relationship to the OSI reference model, as shown in the figure. IEEE 802.2 is a unifying sublayer called Logical Link Control (LLC) which is designed to work above any of the various LAN technologies. LLC is offered either as a connectionless or a connection-oriented service and has a number of similarities with protocols such as High Level Data Link Control (HDLC). LLC will be discussed in more detail in the next chapter. Each of the standardised LAN technologies is then given its own number and deals with the physical layer and the Medium Access Control (MAC) for that technology. These three basic technologies IEEE 802.3 for CSMA/CD, IEE 802.4 for the token passing bus and IEEE 802.5 for the token ring will be described in the following sections.

These IEEE 802 standards have now been adopted as ISO standards and, conveniently, the ISO standard numbers are simply formed by preceding the IEEE number with an extra 8 - so that for example ISO 8802/3 is the international standard for CSMA/CD LANs.

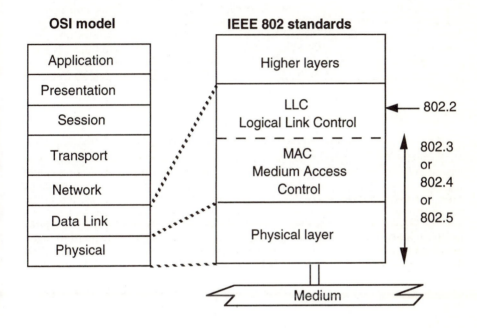

Fig. 7.3 *Relationship between the OSI reference model and IEEE 802 standards*

The relationship between the IEEE 802 standards and the OSI reference model for Open Systems Interconnection is also shown in Fig.7.3. The standards cover the lowest two layers of the OSI model: the physical layer and the data link layer. The data link layer is divided into two sublayers, the lower sublayer defining the medium access mechanism and the higher sublayer being LLC, which is common to all LANs. For some LANs, the Physical layer is also divided into two sublayers dealing with the detail of attachment of system to the network and the the details of the signalling used by the network.

The IEEE 802 committee has gone on to consider a number of other topics for LANs as new techniques have become available, such as the integration of voice and data. They also consider Metropolitan Area Networks, a standard which oscillated for a number of years during its development between those who favoured FDDI, an American standard, and those who favoured the Distributed Queue Dual Bus (DQDB), first specified in Australia. The latter has now been adopted. Both of these network architectures will be discussed in a later chapter. The list of activities in the IEEE 802 committee is given in Table 7.1. (Note that a TAG is a Technical Advisory Group working on a topic which is likely to produce a standard at some time in the future.)

802.1	Relationship between the 802 standards and the OSI model.
802.2	Logical Link Control.
802.3	Carrier Sense Multiple Access with Collision Detection.
802.4	Token-Passing Bus Access Method.
802.5	Token-Passing Ring Access Method.
802.6	Metropolitan Area Networks.
802.7	Broadband TAG.
802.8	Fibre Optic TAG.
802.9	TAG concerned with the integration of LANs and digital telephony.

Table 7.1 *IEEE 802 standards family*

We shall now consider in more detail the three principal standardised LANs: CSMA/CD, token bus and token ring.

7.5   CSMA/CD [IEEE 802.3]

The majority of installed LANs are based on Carrier Sense Multiple Access with Collision Detection. These can range from quite low speeds to 20 Mbit/s and run on a variety of media. A commonly used industry standard is the Ethernet and this is often used as a synonym for CSMA/CD. Other proprietary networks such as Appletalk or the Acorn Econet are also based on the same principle.

The IEEE 802.3 standard can support speeds of between 1 Mbit/s and 20 Mbit/s, but the physical layer standard relates to 10 Mbit/s using a shielded coaxial cable.

7.5.1   MAC service and sublayer

The MAC service to the LLC consists of three primitives:

MA.DATA.request   is used to ask the MAC layer to send a frame onto the network.

MA.DATA.confirm   indicates that the transmission was either successful or unsuccessful.

MA.DATA.indication   signals that an incoming frame has arrived for the LLC layer which is presented together with the source and destination addresses and a status field.

The structure of the MAC frame is shown in Fig. 7.4. The source address is inserted by the MAC layer and the destination address is passed from the LLC in the MA.DATA.request. Although either two or six octets may be used for the addresses, the same convention must be used by all stations on an a single LAN.

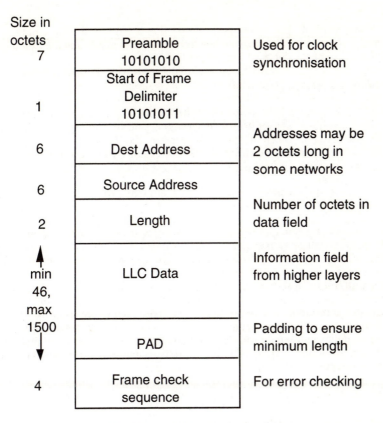

Size in octets		
7	Preamble 10101010	Used for clock synchronisation
1	Start of Frame Delimiter 10101011	
6	Dest Address	Addresses may be 2 octets long in some networks
6	Source Address	
2	Length	Number of octets in data field
min 46, max 1500	LLC Data	Information field from higher layers
	PAD	Padding to ensure minimum length
4	Frame check sequence	For error checking

Fig. 7.4 *CSMA/CD frame format*

The 48-bit address format is shown in Fig. 7.5; the first bit denotes either the address of an individual station or a group address. Group addresses indicate that the frame is destined for *all* of the stations belonging to that group (a logical relationship defined at a higher layer). A special case of the group address is where all the destination bits are set to 1 indicating this frame is broadcast to all of the stations on the network. The second bit indicates whether the addressing scheme used is either globally or locally administered. Similar addressing formats are used for the IEEE 802.4 and IEEE 802.5 standards.

I/G	U/L	46-bit address

I/G = 0 Individual address
I/G = 1 Group address
U/L = 0 Globally administered address
U/L = 1 Locally administered address

Fig. 7.5 *48-bit address format*

The function of the PAD field is to ensure that the frame is long enough for a collision to be heard by all stations on the network. The Frame Check Sequence is a 32-bit cyclic redundancy check.

7.5.2 Medium Access Control

When a MA.DATA.request is issued, the MAC layer constructs a frame in the above format. The medium is sensed for a carrier signal and the station defers to traffic already on the network. When no carrier is detected, the transmission begins and the MAC sublayer provides a serial string of bits for transmission by the physical layer. If transmission is completed successfully, the MAC layer reports this to the LLC using MA.DATA.confirm.

It is possible that more than one station will attempt to transmit at the the same time thus causing a collision by interfering with each other's transmsissions. This can only happen during a "collision window" after which the transmitted frame will have propagated to all stations on the network, which will then defer. If a collision is detected, the transmitting station sends a "jam" signal for sufficient time to ensure that all stations have detected the collision and the transmission is then aborted. The frame will then be sent again after a random time.

The length of time a station must wait before it can be sure that all other stations have deferred is called the *slot time*. This is greater than the round-trip time (i.e. twice the maximum propagation time between any two stations) plus the jam time. After the jam has been issued, a station employs a "truncated binary exponential backoff" procedure to determine when to retransmit. At the nth retransmission attempt, the station delays a multiple, r, of the slot time, where r is a uniformly distributed random integer in the range

$$0 \leq r \leq 2^k$$

where $k = \min(n,10)$

This procedure ensures that the network is used efficiently if lightly loaded but reduces the number of collisions due to retransmissions when the network is heavily loaded. After a maximum number of attempts, an error is reported. For the 10 Mbit/s implementation with appropriate baseband coaxial cable, the parameter values for CSMA/CD are:

    slot time = 512 bit times
    max. number of retransmission attempts = 16
    jam size = 32 bits
    maximum LLC data field = 1500 octets
    minimum LLC data field plus PAD = 46 octets

Stations on the network may expect to receive frames at any time and examine the destination address of all the frames and, if this matches their own address or is a group to which this station belongs or is the broadcast address, the frame will be copied from the network provided it has not suffered a collision.

7.5.3 Physical layer and size constraints

CSMA/CD physical components are shown in Fig. 7.6. The standard medium attachment unit (transceiver) comes with a pin which, when clamped into place, makes the appropriate electrical connection with the coaxial cable. The transceiver is then attached to the controller board in the station by an Attachment Unit Interface (AUI) cable which can be up to to 50 m in length. This cable has four twisted pair circuits offering separate data and control circuits in each direction. The circuits are independently clocked and the data circuits use Manchester encoding.

The topology of a CSMA.CD network is in the form of a tree with repeaters connecting individual segments (shown in Fig. 7.6). The repeaters connect two transceivers and repeat all the signals from one segment to the other, providing appropriate amplification and retiming. The maximum segment length is 500 m with a maximum of 100 transceivers per segment. The maximum distance between any two stations using this topology is 2.5 km.

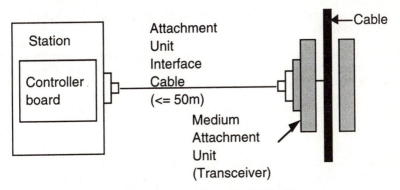

Usual attachment to coaxial cable

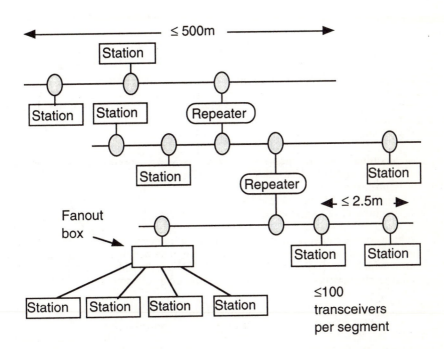

Fig. 7.6 *Medium attachment and typical CSMA/CD configuration*

A common method of attaching a group of stations is by a "fanout box" with one transceiver and AUI cables to up to 8 stations. A cheaper version of CSMA/CD is also available (called Cheapernet). This uses thinner, more flexible cable which can be taken directly into a station's circuit board. It also operates at 10 Mbit/s but is restricted to a maximum segment length of 185 m.

7.6 Token passing bus [IEEE 802.4]

The second of the standard LAN technologies we consider is a token passing system based on the broadcast bus. The token passing scheme allows maximum use to be made of the available bandwidth even when the network is busy. Its bus topology is suitable for situations such as industrial production lines. For these reasons it was adopted as the network for the Manufacturers' Automation Protocol (MAP) which recommends a suite of protocols at all layers of the OSI stack for the manufacturing industry.

7.6.1 Medium access and ring maintenance

The medium access and maintenance of the token passing mechanism is done in a distributed way with all stations contributing to the procedure. A logical ring of stations is formed as illustrated in Fig. 7.7. Inactive stations will not be a part of this ring, but may choose to listen to the transmissions of other stations.

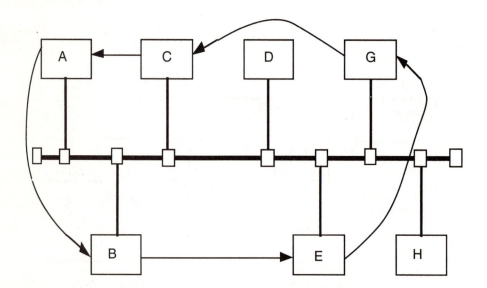

Fig. 7.7 *Logical ring of stations on token bus*

A single token (illustrated in Fig 7.8) is passed in turn by each station to its successor in the logical ring. The Destination Address field is set to the station address of the successor. Possession of the token permits a station to transmit one or more data frames onto the network. Having done this, it must then send the token on to the next station. The data frame (also illustrated in Fig 7.8) may be marked as a "Request for Response" frame, indicating that an immediate response is required from the recipient. The recipient is then allowed to transmit a response frame even though it does not hold the token. The number of frames which a station may transmit before releasing the token is bounded by the need to keep within an agreed token rotation time. All stations measure the actual rotation time and use this as a guide to how much they can transmit. It is also possible to have up to eight priority levels with higher priority frames being transmitted first whilst always keeping within the token rotation time.

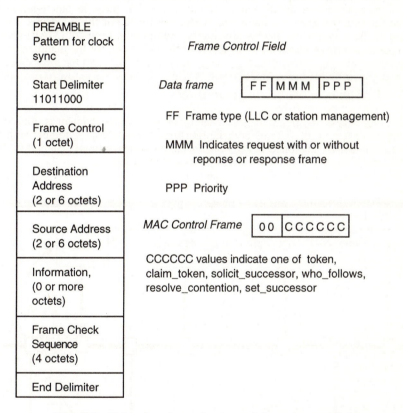

Fig. 7.8 *Token bus frame format and frame control fields*

It is essential that the logical ring is maintained: that stations wishing to join or leave the ring can do so and that recovery action is taken in the case of a lost or duplicated token. There are therefore a number of other frame types which enable the stations to take part in a ring maintenance procedure as part of the MAC layer protocol.

Immediately after sending a token, the station listens to check that its successor heard the token and is still active; this will be indicated when the successor station sends a valid frame (data, token or other) onto the network. If nothing is heard, the token holding station sends the token again and listens again. If nothing is heard the second time, it is assumed that the successor station has failed. The token holding station sends a "who_follows" frame containing its successor's address . All stations check whether their predecessor address matches the address indicated. The station which matches this value becomes the new successor to the token holding station and responds to the "who_follows" with a "set_successor" frame. The token holding station can then transmit the token to its new successor and the failed station is excluded from the logical ring. If no "set_successor" is forthcoming, the token holding station sends a "solicit_successor" frame to find another station prepared to act as its successor.

Stations are added to the ring by means of the "solicit_successor" frame which is sent by active stations. This frame specifies a range of potential addresses as successors. (Stations are always included in the logical ring in descending station address order except for the station which completes the ring.) A station wishing to enter the ring can then respond with a "set_successor" frame. Contention from more than one potential successor is resolved with a "resolve_contention" frame which imposes a time order on the responses. Initialisation of the ring is done by successively adding new stations to the ring.

A station wishing to leave the logical ring simply does not send the token on; this will be detected by the token holding station which initiates the recovery procedure already discussed.

7.6.2 The physical layer

There are three physical layer options quoted in the standard. These are;

(1) An omnidirectional bus using phase continuous frequency shift keying at 1 Mbit/s

(2) An omnidirectional bus using phase continuous frequency shift keying at 5 or 10 Mbit/s.

(3) A directional bus with active head end repeater using multilevel duobinary amplitude modulation phase shift keying at 1, 5 or 10 Mbit/s.

7.7 Token ring [IEEE 802.5]

The token ring was originally developed by IBM as its own LAN network architecture. Because of its likely widespread use, it was adopted as one of the standard IEEE standards. The standard originally specified a speed of 1 or 4 Mbit /s but now an option of 16 Mbit/s is available. Signals are transmitted using differential Manchester encoding.

Stations are connected together serially in the form of a ring, with information being transmitted sequentially, bit by bit from each station to the next station which regenerates and repeats each bit, and can also read (copy the bit) or write (change the bit) as it passes

7.7.1 Medium access

A single token circulates around the ring giving each station in turn permission to transmit information frame(s). A maximum token holding time limits the number of frames which can be sent before the token must be passed on, ensuring that no station monopolises the network. A transmitted frame circulates around the ring and is copied by the destination station(s). The frame is removed from the ring by the sending station. This:

(1) gives the receiving station an opportunity to indicate its response to the frame,
(2) ensures fair access to the ring by always passing the token in strict sequence around the ring.

The formats of the token and of the information frame are shown in Fig. 7.9. The source and destination addresses use a similar format to both the CSMA/CD and the token bus LANs.

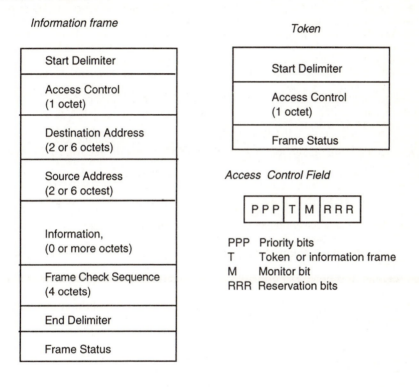

Fig. 7.9 *Token ring frame formats and access control field*

The frame status field consists of two bits. If a station recognises its own address in the destination address field of the frame it sets the "Address Recognised" bit to 1. Additionally, if the information frame is copied by the receiving station, the "Frame Copied" bit is set to 1. These bits can then be inspected by the transmitting station when the ring returns.

7.7.2 The monitor station

The majority of maintenance functions on the token ring are performed by a monitor station, whose job is to provide normal operation and to recover from error conditions. The single active monitor station regularly broadcasts its presence by issuing "Active Monitor Present" frames. The first station that receives this frame starts a timer and sends a "Standby Monitor Present" frame. Expiry of this timer before another indication is heard from the active monitor station indicates that the active monitor has failed, at which point the standby monitor takes over.

The active monitor is responsible for the detection of various error conditions and will attempt to correct these and return the ring to a working state. It is also responsible for providing a variable length delay-buffer which is included in the ring to ensure the ring is greater than its minimum length of 24 bits (the number of bits in the token). The monitor also provides the clock to which all other ring stations synchronise. The monitor bit in all frames (see Fig 7.9) is set to 0 by the transmitter of the frame. The monitor converts the bit to 1 as the frame passes. If a frame of priority greater than 0 with the monitor bit set to 1 arrives at the monitor, it can be assumed that the frame has attempted to circulate twice round the ring and it is aborted by the monitor. A token with priority 0 is allowed to circulate as this indicates that the ring is idle.

7.7.3   Priority mechanism

The token ring allows up to 7 priority levels. The current ring priority is contained in the PPP bits of the token (see Fig 7.9). This allows any station to transmit frames at this or a higher priority. The priority of the ring is raised by a station which receives a token with the request filed, RRR, set higher than the current priority. It does this by changing the token's PPP bits to the new higher value and setting the RRR bits to 0. At this point, the station becomes a *stacking station* and remembers both the old and new priority levels by pushing them down onto two stacks. This must be done because, in order to offer fair access to all stations at any priority level, the station which increases the priority is also responsible for resetting it to its original value when there is no longer a requirement for the higher priority. At this stage, both priority values are popped from the stack and the original priority level is inserted into the token. Both values are stacked because they are needed since the same station may have to raise the priority more than one time before it is possible to return to the original level (e.g. from 1 to 3 and later from 5 to 6). A full description of the priority mechanism can be found in the standard.

7.8   Conclusion

The three standards just discussed are all reasonable candidates for the selection of a LAN. In performance terms, the token systems are more predictable and offer high utilisation at high loads (when less time is spent passing tokens). At low loads, the token passing, particularly on the token bus, takes up a noticeable proportion of the capacity. The CSMA/CD technique is very efficient and offers low delay with light traffic, but, as the load is increased, the delays become longer and less predictable. For those interested in the details of LAN performance the book by Hammond and O'Reilly is recommended [9]. Most installed networks tend to run at low utilisation and it is a reasonably simple matter to install a bridge to partition a single LAN into two if overload becomes a problem.

The token passing systems also offer a range of priorities and a bound on the delay time. Both of these aspects are important when consideration is given to carrying traffic such as real-time speech on a LAN. Regardless of technical comparisons, the choice of LAN is often decided by the most prevalent computing systems attached to the LAN, which will, de facto, offer simpler or cheaper access to one particular LAN type.

7.9   References

1.   Metcalfe, R.M. and Boggs, D.R., "Ethernet: Distributed Packet Switching for Local Computer Networks" *Communications of the ACM,* 1976, **19** no 7, pp. 395-404

2.   Wilkes, M.V. and Wheeler, D.J., 1979, "The Cambridge Digital Communication Ring" Local Area Communications Network Symposium, Boston, USA, pp. 47-60, Mitre Corp

3.   Abramson, N., "The ALOHA system - another alternative for computer communications", Proc AFIPS Fall\Joint Computing Conf., Houston, Texas, pp 281-285, 1970

4.   IEEE 802.1, Relationship of the IEEE 802 standards with the ISO OSI Model, American National Standards Institute

5.   IEEE 802.2, Logical Link Control Specifications. American National Standards Institute, (1985)

6.   IEEE 802.3, Carrier Sense Multiple Access with Collision Detection (CSMA/CD) Access Method and Physical Layer Specification. American National Standards Institute, (1985)

7.   IEEE 802.4, Token-Passing Bus Access Method and Physical Layer Specifications. American National Standards Institute, (1985)

8.   IEEE 802.5, Token Ring Access Method and Physical Layer Specifications. American National Standards Institute, (1985)

9.   Hammond, J.L. and O'Reilly, P.J., 1986, "Performance Analysis of Local Computer Networks" (Addison Wesley)

Chapter 8

LANS: Protocols above the medium access layer

A. Gill Waters

8.1 Introduction

In this chapter, we examine protocols for LANs above the Medium Access Layer. The IEEE 802.2 standard for Logical Link Control is first discussed. This is followed by a very commonly used suite of protocols, those designed for the Internet. The Internet is a large and growing collection of interconnected networks, which has its origins in the provision of packet switched networks to the academic and military community in the USA. The suite is offered as the standard communication package for many workstations and personal computers often over CSMA/CD networks and it has become a de facto standard. The protocols are sometimes called the TCP/IP protocols as they include the Internet Protocol (IP) which offers a standard datagram format and Transmission Control Protocol (TCP) which offers reliable bidirectional streams of information.

The final two sections of the chapter look briefly at the ways in which LANs can be connected together and at proprietary personal computer LANs.

8.2 Logical Link Control   [IEEE 802.2]

Logical Link Control (LLC) is the upper sublayer of the data link layer which is the standard for all the IEEE 802 LAN technologies discussed in the previous chapter (see Fig. 8.1). It enables processes in one station to send and receive frames across the LAN to one or more processes in other station(s) and it also offers multiplexing of several process-to-process conversations. This is done in a uniform manner regardless of the underlying LAN type.

The standard includes three principal elements: the Logical Link Service, offered to higher layers by the LLC, the LLC protocol (the rules of procedure used between peer LLC implementations to achieve this service) and the service which is required of the next lower sublayer (the Medium Access sublayer).

8.2.1   LLC services

Three types of service are offered - connectionless, connection-oriented and an acknowledged connectionless service. The connectionless service has minimum protocol complexity. It enables single data frames to be sent from a process in one machine to process(es) on other machine(s) across the network without the need for a pre-established connection. The frame transmission may be point-to-point, multicast or broadcast. The connection-oriented service supports point-to-point connections down which data can be sent and includes provision for sequence numbers, flow control and error control. The acknowledged connectionless service is for Request /Response transactions or status polling.

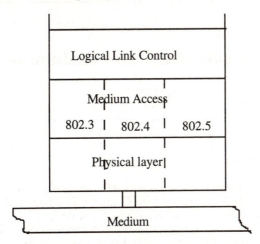

Fig.8.1 *Relationship of LLC to other IEEE 802 LAN protocols*

Connectionless services

L_DATA.request	Transmit frame
L_DATA.indication	Incoming frame has arrived

Connection-oriented services

L_CONNECT.request	Establish a logical link connection
L_CONNECT.indication	
L_CONNECT.confirm	
L_DATA_CONNECT.request	Send information frame on
L_DATA_CONNECT.indication	connection
L_DATA_CONNECT.confirm	
L_DISCONNECT.request	Terminate connection
L_DISCONNECT.indication	
L_DISCONNECT.confirm	
L_RESET.request	Reset connection to its initial state
L_RESET.indication	
L_RESET.confirm	
L_CONNECTION_FLOWCONTROL.request	Indicates amount of data that can
L_CONNECTION_FLOWCONTROL.indication	be passed before acknowledgement

Table 8.1 *LLC services*

A list of the LLC service primitives for the unacknowledged connectionless service and the connection-oriented service is shown in Table 8.1. A *request* is made by a service user; this is presented as an *indication* to the remote service user; the *confirm* indicates that the required action has been completed by the LLC. These terms are frequently used in computer network standards.

8.2.2 LLC Protocol

The LPDU (LLC Protocol Data Unit) is shown in Fig. 8.2. LLC is based on HDLC LAP-B (High Level Data Link Control, Balanced) data link control procedures which form part of the CCITT's X.25 recommendations for connection to a public packet switched network [1]; a number of extra frame types have been added. The Source and Destination Service Access Point Addresses (DSAP and SSAP) are used to identify individual processes (instances of applications). Eight bits are allocated to each field; bit 0 in the Destination SAP indicates whether a group or individual address is identified. Bit 0 in the Source SAP is used to indicate if the frame is a Command or Response frame (c.f. HDLC terminology).

DSAP address (8 bits)	SSAP address (8 bits)	Control field (8 or 16 bits)	Information (Zero or more octets)

Note: The upper bound for M is a function of the Medium Access control technique used.

Fig. 8.2 *LLC PDU format*

Information and Supervisory frames for connection-oriented service		
I	Information	LLC Data
RR	Reset Ready	}
RNR	Reset Not Ready	} Flow and error control
REJ	Reject	}

Un-numbered frames		
SABME	Set Asynchronous Balanced Mode	Establish data link connection
DISC	Disconnect	Terminate data link connection
UA	Un-numbered acknowledge	Acknowledge SABME or DISC
UI	Un-numbered information	Connectionless frame
XID	Exchange ID	Notify LLC services supported
TEST		Test LLC-LLC interconnectivity
FRMR	Frame Reject	Irrecoverable error in frame
AC0	Acknowledgements	Replies for acknowledged
AC1		connectionless service

Table 8.2 *LPDU frame types*

The Control field is 16 bits long for frames containing sequence numbers (data and flow control (supervisory) frames for the connection-oriented service), offering 7-bit sequence numbers. Un-numbered frames have 8-bit control fields. The full list of frame types is given in Table 8.2. For the connectionless service, three frame types have been introduced: UI for connectionless data, XID which allows LLC implementations to exchange information on their capabilities, TEST which tests that communication exists between LLC implementations and two acknowledgement frame types for the acknowledged connectionless service.

8.3   The Internet protocols

The Internet protocol suite, supported by the USA Department of Defense (DoD), has a long history. The Defense Advanced Projects Research Agency (DARPA) Internet is a large collection of interconnected networks of many types including various LANs and WANs and terrestrial, satellite and radio links. It has evolved from a number of separate networks, principally the Arpanet, a research network which grew rapidly during the 1970s and 1980s. The Internet connects academic, military and other organisations and now extends beyond the USA to include countries such as Australia and is directly connected to many other networks internationally.

One reason for the importance of the protocol suite is that it must be supported by any equipment attached to the Internet. This has put pressure on computer suppliers to offer the protocols, which are now available for a large range of computers from personal computers and workstations to mainframes. Recently, the UK JANET network (Joint Academic NETwork) has added the Internet suite to its range of protocols and it is now possible to communicate directly from systems attached to JANET to machines on the Internet.

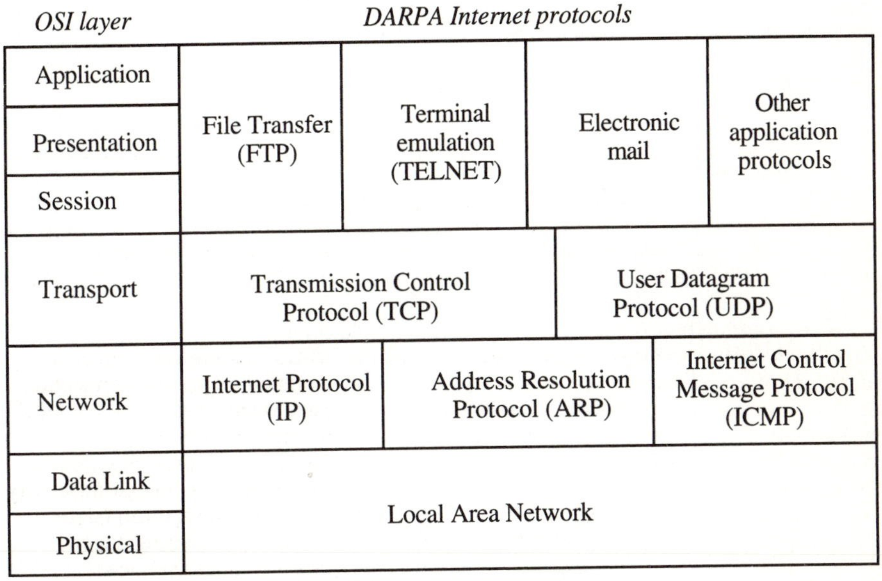

OSI layer	*DARPA Internet protocols*			
Application	File Transfer (FTP)	Terminal emulation (TELNET)	Electronic mail	Other application protocols
Presentation	File Transfer (FTP)	Terminal emulation (TELNET)	Electronic mail	Other application protocols
Session	File Transfer (FTP)	Terminal emulation (TELNET)	Electronic mail	Other application protocols
Transport	Transmission Control Protocol (TCP)		User Datagram Protocol (UDP)	
Network	Internet Protocol (IP)	Address Resolution Protocol (ARP)	Internet Control Message Protocol (ICMP)	
Data Link	Local Area Network			
Physical	Local Area Network			

Fig. 8.3   *The Internet protocol suite*

The protocol suite is designed to ease communication between different types of networks and is based on the datagram (connectionless) approach. The various elements of the protocol suite are shown in Fig. 8.3, which also shows the rough correspondence with the OSI reference model. (Note that the figure shows examples of facilities at each layer but does not imply direct vertical relationships between individual components in the Internet suite.)

The details of the Internet or TCP/IP suite and a number of associated documents are published by the DoD and are freely available across the Internet. They are part of a collection of Request for Comments (RFCs) which also cover many other networking topics. A list of the principal RFCs related to the TCP/IP protocol suite is included in the reference section at the end of this chapter.

The Internet suite runs on many different LANs. It is possible to mount IP directly on top of the MAC layer of a LAN, but the current recommendation is to use the IEEE 802.2 LLC connectionless service where possible, using the Un-numbered Information frames and specified values for DSAP and SSAP addresses etc. (The details are given in RFC 1042.)

At the network layer, the Internet Protocol (IP) enables individual datagrams to be sent across the network. It offers a number of addressing options and message fragmentation and reassembly. Also at this layer are the Address Resolution Protocol (ARP) [RFC 826] whose job is to map Internet addresses to Ethernet (CSMA/CD) addresses. ARP uses broadcast transmission to advertise the system's own address and to request information from other systems on the network. The Internet Control Message Protocol (ICMP) [RFC 792] enables IP entities to exchange control and maintenance information.

The transport layer consists of two sets of services accessed directly to the application. The Transmission Control Protocol (TCP) offers bidirectional reliable byte streams and employs error detection and correction and flow control procedures. The User Datagram Protocol (UDP) offers a connectionless service at this layer.

A number of standard applications are offered which use either TCP or UDP. These include file transfer, electronic mail etc. We shall now look at some of the elements of the Internet protocol suite in more detail.

8.3.1 The Internet Protocol (IP)   [RFC 791]

The IP protocol is best described by reference to the IP header (Fig. 8.4) which is prepended to each data unit to be sent across the network (Fig. 8.5 shows the position of the IP header related to the rest of the protocol hierarchy.)

There are length fields for both the header (in 32-bit words) and the complete frame, including the header (in octets). The type of service consists of flags to specify reliability, precedence, delay and throughput parameters. The identification field is a unique value for this datagram.

A message which is too long to fit into a single data unit is fragmented and the offset indicates where this fragment starts in relation to the beginning of the message (in 64-bit units). The time to live file is decreased as an IP datagram travels across a network boundary; if the time to live has expired, the datagram is discarded. This prevents datagrams from circulating endlessly in the Internet. The protocol field identifies the protocol of the next higher layer (TCP or UDP). The checksum offers early identification of header errors; it is recomputed at each gateway because the time to live field is updated. IP is an unreliable protocol so a checksum error leads to the datagram being discarded. The source and destination addresses will be discussed in the next paragraph. The options field is variable length and indicates required facilities such as a specified route to be taken. Padding ensures that the IP header falls on a 32-bit boundary. The data field is a multiple of 8 bits with a limit of 65,535 octets for the whole of the IP datagram (including the header).

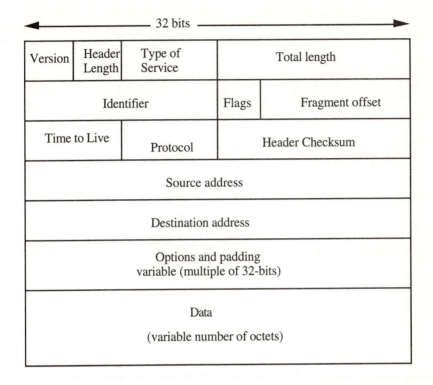

Fig. 8.4 *Internet Protocol header*

MAC layer header	IP header	TCP header	Application Information	CRC

≥ 20 octets   ≥ 20 octets   e.g. FTP

Fig. 8.5 *Nested Internet protocol headers in a MAC layer frame*

8.3.2 IP addressing

A number of different addressing classes are offered to suit different sizes of networks. The class is indicated by the first few bits of the address as shown in Fig. 8.6. A multicast address option is also specified, enabling an IP protocol unit to be sent to all the members of a multicast group.

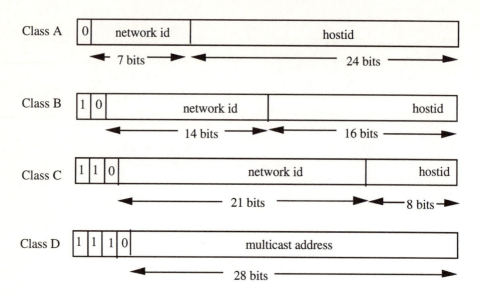

Fig. 8.6 *Internet address formats*

8.3.3 Transmission Control Protocol (TCP)   [RFC 793]

TCP offers reliable communication on top of the datagram service offered by IP. It includes procedures to set up and clear down a connection and to send segments of up to 64 kbytes in both directions. The connection set-up procedure is undertaken using a three-way handshake using agreed combinations of the SYN and ACK control flags in the TCP header (Fig. 8.7). This enables each end first to know that the connection is agreed and, secondly, to indicate the initial window size and the first sequence number to be sent. The port numbers provide the addressing needed to determine the actual application process within a machine which is being used (e.g. an instance of an FTP transfer or of a mail message transfer), thus allowing concurrent support for several processes. A connection is set up between a source port at one machine addressed by the source IP addresses and a destination port within the machine at the IP destination address. Well-known port numbers are generally assigned for the "Listening" end of a specific service, e.g. FTP. A connection is cleared by setting the FIN flag in the control field.

Once the connection has been established, data bytes may be sent. TCP offers a reliable stream in each direction, providing error detection and recovery and ensuring that the information arrives in the correct order. The sequence number indicates the byte position of the first octet of data relative to the beginning of the segment. The acknowledgement number, which is piggybacked onto packets carrying data, acknowledges receipt (by the TCP entity) of all of the octets preceding the value of the acknowledgement number. Several packets may be in flight at any one time thus making efficient use of the networks(s). The recipient also specifies the window size in the Window field which is the number of octets for which it has buffering available. Error recovery is done by the transmitter timing out and resending information for which an expected acknowledgement has not been received. In general, TCP attempts to fill packets when it can, so as not to waste capacity by sending short packets. If a short message must be sent quickly (e.g. where a simple command has been typed at a keyboard) the Push function is used to send the information immediately. A priority message, which does not obey normal flow control procedures, can be sent using the Urgent bit in the control field, with the Urgent Pointer indicating the first octet in the segment which follows this urgent data.

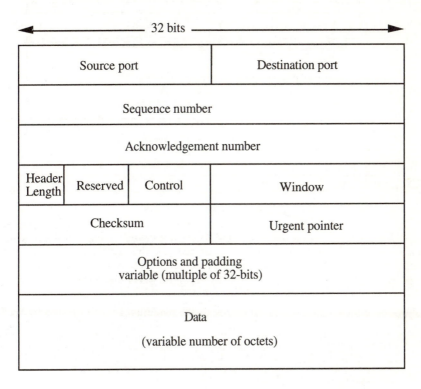

Fig. 8.7   *TCP protocol header*

8.3.4 User Datagram Protocol (UDP)   [RFC 768]

In contrast to TCP, the User Datagram Protocol is very simple. This allows unreliable transmission using the IP layer. The UDP header is shown in Fig. 8.8.

The source and destination port, as for TCP, specify the actual process which is sending (or receiving) the datagram. The other fields specify the length of the UDP datagram and the checksum provides error detection, but UDP does not undertake error recovery.

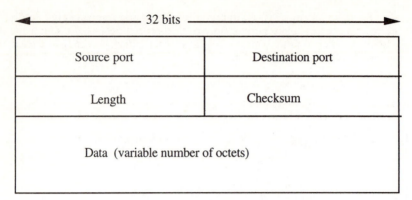

Fig. 8.8 *UDP Protocol header*

8.3.5 *Application layer protocols*

A number of standard application protocols are defined and are used widely on the Internet. The *Telnet* protocol allows users to use their local terminal to log on remotely to another machine. Telnet uses TCP and offers a *network virtual terminal* which enables the user's real terminal to be described and thus understood by the host system. The *FTP* protocol enables suitably authenticated users to move through a remote file system, to examine the contents of directories and to send files to and receive files from the remote system. It uses two separate TCP connections: one for control messages and one for the data transfer. Electronic mail uses the Simple Mail Transfer Protocol (*SMTP*) which again makes use of the reliability offered by TCP. If the remote machine specified as the destination is not available, the message remains on the sending host until it can be transferred. In addition to these standard protocols, most implementations allow user code to be written at the application layer and to use either UDP or TCP to implement their own communication application.

8.4 Bridging and routing

It is frequently necessary to interconnect LANs and to connect LANs to Wide Area Networks (WANs) in order to gain access to other resources or to communicate with other people. Fig. 8.9 shows some of the more common interconnection units which must be installed to achieve this interconnectivity. Sections of Local Area Networks which are connected at the physical layer are called *repeaters*; they simply regenerate and transmit the electrical signals on to the next network segment. Systems interconnecting LANs are generally called *bridges* and are connected at Layer 2; these systems need appropriate interfaces to both of the networks they connect. Bridges can be reasonably simple to implement as LANs share a common protocol at the LLC layer. Where LANs and WANs are interconnected, the mapping is done at layer 3 of the OSI model; these systems are known as *gateways* (the European term) or *routers* (the American term). They must take account of the many possible differences between the two different networks, which will include connection-oriented or connectionless working, different maximum packet size, different addressing schemes etc. The term router reflects the fact that systems such as these must be able to determine where to send the information whether the destination system lies within one of the directly connected networks or on a network which is reached via one of them.

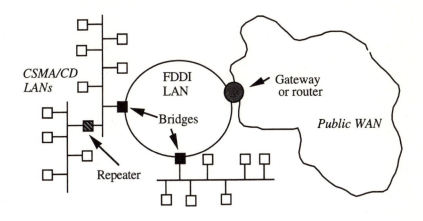

Fig. 8.9 *Typical interconnection systems*

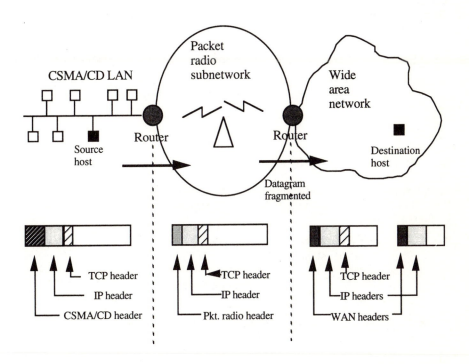

Fig. 8.10 *Encapsulation of IP datagrams at network routers*

Use of the TCP/IP suite makes the functions of bridges and routers reasonably simple, once an appropriate route is known. The reliability achieved by TCP is end-to-end, so the routers need not be concerned with error recovery or flow control procedures. Fragmentation may be necessary when a datagram enters a network with a smaller maximum packet size than the one it left. In order to traverse a particular network, the IP packet will be encapsulated in a protocol frame for each network as it enters the network and unwrapped by the router as it leaves the network. Both encapsulation and fragmentation are shown in Fig. 8.10.

8.4.1 Bridges

Bridges are reasonably simple; they forward frames as quickly as possible and provide buffering for where this is not immediately possible. If buffers overflow because congestion has occurred, frames may be discarded. Unfortunately, each IEEE 802 LAN standard has a different frame format, so translation must be done by the bridge. They also have different data rates making congestion more likely when forwarding, for example, from a CSMA/CD network operating at 10 Mbit/s to a token ring operating at 4 Mbit/s. Their difference in frame lengths is serious as the MAC protocols do not offer segmentation and reassembly, so in some cases frames which are too long must be discarded.

There are two principal routing strategies for bridges [8]. The first is the *transparent bridge*, which gradually learns the route to distant hosts. Initially all frames are flooded onto all possible connected networks. As each frame arrives its source address indicates where a particular host is situated, so that the bridge learns which way to forward future frames to that address. To prevent looping, the bridges form a spanning tree, connecting all bridges and offering a single route between bridges. In the *source routing bridge*, users must specify which route is to be followed. A route is found by sending a discovery frame which is replicated to use all possible paths to the destination; the frame collects bridge addresses as it passes through them. The destination responds to each discovery frame, enabling the source to choose the most appropriate route. A comparison of the two techniques can be found in [9].

8.5   Personal Computer networks

In addition to the standard network technologies and protocols described in the IEEE 802 standards, there are a number of proprietary approaches to LAN interconnectivity. Examples include:

> NETBIOS
> Novell's Netware
> 3-Com Networks
> The Macintosh network system, Appletalk
> Acorn Computers' Econet

IBM's NETwork Basic Input/Output system (*NETBIOS*) aims to provide a fixed interface between the operating system and the LAN which is independent of the network hardware or software. NETBIOS can be seen as an interface between the Presentation and Session Layers of the OSI reference model and it offers both virtual circuits and datagram services. An application communicates with NETBIOS by issuing commands in the form of a data structure called a Network Control Block. Some PC token ring interface cards use an on-board microprocessor which relieves the host PC from the protocol processing load of both the ring protocol and NETBIOS.

Novell's *Netware* is a portable Networking Operating System (a system which allows both local and remote processing, often transparently, to the user), with implementations for many different LAN types. Netware is constructed by inserting a Netware shell, a piece of code between the application and the PC operating system which intercepts commands and decides whether or not they need to communicate with other systems. The communication package is called Internetwork Packet Exchange (IPX), an interworking protocol with similar facilities to TCP/IP. Netware includes password authentication for users and fault tolerant features based on duplication of filestores.

8.6   Conclusion

This chapter has given an introduction to some of the more important protocol suites commonly used by systems attached to LANs. The Internet suite has significance for many networks, not just LANs. Further information on the Internet suite can be obtained from the RFCs or from books such as that by Comer [10]. Network interconnection is an important issue, which has only been touched on briefly and to which later chapters will return. Further reading could start with Chapter 9 of [11] or the book by Miller [12].

8.7   References

1.   CCITT Recommendation X.25, Vol VIII, Facsicle VIII.3, International Telecommunication Union, Geneva, 1988
2.   RFC 768: Postel, J.B., "User datagram Protocol", August 1980
3.   RFC 791: Postel, J.B., "Internet Control message Protocol", September 1981
4.   RFC 792: Postel, J.B., "Internet Protocol", September 1981
5.   RFC 793: Postel, J.B., "Transmission Control Protocol", September 1981
6.   RFC 826: Plummer, D.C.,   "Ethernet address resolution protocol for converting network protocol addresses to 48-bit Ethernet addresses for transmission on Ethernet hardware", November 1982
7.   RFC 1042: Postel, J.B. and Reynolds J. K.," Standard for the transmission of IP datagrams over IEEE 802 networks", February 1988
8.   *IEEE Network,* 1988: Special issue on bridges and routers, **2** no. 1
9.   Soha, M. and Perlman, R., "Comparison of two LAN bridge approaches", *IEEE Network,* 1988, **2** no. 1, pp. 37-43
10.   Comer, D. E., 1988, "Interworking with TCP/IP: Principles, protocols and architecture" (Prentice Hall)
11.   Waters, G. (ed), 1991, "Computer Communication Networks" (McGraw Hill)
12.   Miller, M. A., 1991, "Internetworking: a guide to network communications LAN to LAN; LAN to WAN" (M& T Publishing)

Chapter 9

Wide area networking

Keith Caves

9.1 Introduction

In this chapter, we shall discuss the support of inter-site corporate data communications across the wide area.

An introductory section sets the scene by indicating the trends that are creating the requirements for increased performance from managed data services in wide area networks.

Next, we see how the traditional vehicle for inter-site networking - the leased line - is failing to meet the growing needs for performance and management at affordable costs.

Two services that are designed to meet the above needs are then discussed. The first - frame relay - provides an HDLC-based inter-LAN communications service. The second - switched multimegabit data service - is a high speed connectionless data service that anticipates the B-ISDN.

Finally, the concluding section shows how data internetworking can be provided by both frame relay and connectionless data services in a B-ISDN/ATM environment.

9.2 Internetworking requirements

Over the last few years, important changes have been taking place in corporate data communications networking. The advent of low cost personal computers and workstations has ensured that processing power is available on every desk that needs it. The developments in personal computing have created a need for a new networking environment to enable communications between devices for the support of distributed, interactive applications as well as for ease of access to common equipment such as printers and servers. This new networking environment is based on local area networks (LANs) and peer-to-peer protocols; LANs are increasingly replacing the traditional, centralised, transaction-oriented data communications networks that rely on hierarchical protocols.

LANs use simple, efficient HDLC-based (packet) protocols that ensure high data transfer rates and low latency. Typical applications include file transfer and graphic image transfer, often involving megabytes of data. However, with the data rates available on modern LANs (e.g. up to 100 Mbit/s for FDDI), short intra-LAN response times can be assured. The challenge now is to ensure satisfactory response times for inter-LAN communications involving transmission across wide area networks (WANs).

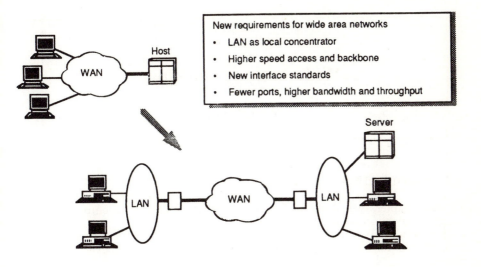

Fig. 9.1 *The emerging corporate data networking model*

The transition towards this new corporate data networking model is illustrated in Fig. 9.1. As shown, the LANs act as local concentrators for traffic that needs to traverse the WAN. A LAN requires fewer physical access ports to the WAN than the hierarchical model, but with higher bandwidth and throughput per port. The explosive growth in LANs poses requirements for new WAN interface standards as well as for network management functionality capable of controlling the dynamic, distributed processing environment. Above all, new WAN services are required which can efficiently support the connectionless protocols used within the local area.

9.3   Leased lines

LAN protocols are designed to ensure the efficient transfer of large amounts of data with short response times, in a shared medium environment. However, transfer of LAN traffic across the wide area is usually transported via leased lines which are generally restricted to either 64 kbit/s or 2 Mbit/s transmission rates. Table 9.1 shows how the transfer time for different types of data files is affected by the transmission rate on the medium.

What the table shows is that, even with a WAN link running at 2 Mbit/s, the transfer time for certain types of inter-LAN traffic across the wide area is barely acceptable.

File Type	Network Type	Transfer Time			
		WAN		LAN	
		64 kbit/s	2 Mbit/s	10 mbit/s	100 Mbit/s
Page of text		170 msec	6 msec	1.2 msec	120 μsec
Spreadsheet		6 sec	200 msec	40 msec	4 msec
Graphics image		15 sec	500 msec	100 msec	10 msec
Large file		1 min	2 sec	400 msec	40 msec

Table 9.1 *File transfer time as a function of bit rate*

Currently, corporate networks which require to interconnect LANs across the wide area typically use leased lines at either 64 kbit/s or 2 Mbit/s. The simplest form of leased line network topology for the interconnection of a number of corporate sites is the mesh - this is illustrated in Fig. 9.2. The drawback with this simple topology is that, as the number of sites begins to grow, so the number of leased lines required for their interconnection increases dramatically. In North America, the average number of sites per corporate customer currently being networked across the wide area for inter-LAN communications exceeds 12, for which a fully connected mesh network would need 66 leased lines. Clearly, few corporations would be happy with the level of costs and the network management problems associated with a structure of this type. Steps are normally taken to reduce the level of connectivity so that only the major corporate sites are mesh-connected with the smaller sites star-wired to their nearest major site.

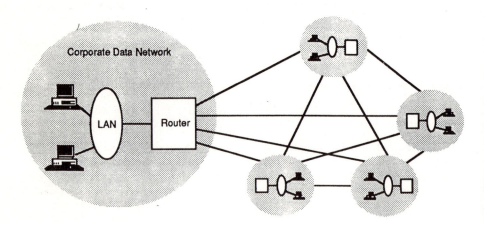

Fig. 9.2 *Corporate networking with leased lines*

Notwithstanding these measures, networks consisting of 20 leased lines at 2 Mbit/s are not uncommon. In the UK, the annual rental charges for such a network might typically average out at around 850,000 ECU. In continental Europe, leased line costs are much higher than in the UK. In Germany, for example, a similar network could easily incur annual charges of 2.9 million ECU whilst, in Italy, the figure would be 4.7 million ECU. Not surprisingly, the high level of costs for leased bandwidth in Europe is a major deterrent for corporate customers.

The cost factor is exacerbated when the characteristics of the traffic on inter-LAN links are considered. Generally, these links are used to carry short, infrequent bursts of data, leading to poor utilisation of available bandwidth and high per-packet data transport costs. Pressure is thus exerted to lease the lowest cost (i.e. minimum bandwidth) lines, but this is incompatible with the requirement for low transport delay. The result is usually a compromise which is perceived by the customer to provide unsatisfactory service at too high a cost.

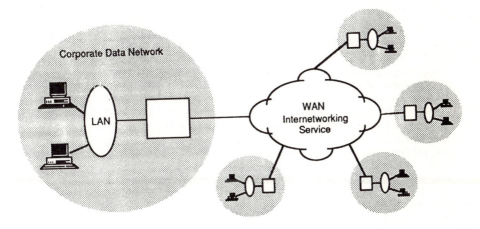

Fig. 9.3 *Point-to-multipoint LAN interconnect service*

Clearly, the characteristics of inter-LAN traffic, when coupled with its high bandwidth and low transport delay requirements, are not well suited to an affordable point-to-point leased line network topology. Ideally, what is needed is an internetworking service which enables traffic from many different sources to be interleaved over a single communications link, to provide cost-effective point-to-multipoint network connectivity. This is illustrated in Fig. 9.3, in which the mesh-connected leased line network has been replaced by a WAN service 'cloud' providing point-to-multipoint communications.

The type of service embodied in the figure has existed for many years, in the form of the X.25 packet switching service. Unfortunately, X.25 suffers from a number of drawbacks. In particular, the protocol was designed for use in error-prone environments and carries substantial associated overheads. It is considered rather a 'heavyweight' protocol for modern, digital, low-error transmission systems and, as a consequence, requires a lot of protocol processing at network nodes. This in turn severely limits the throughput obtainable, causes transport delays and effectively prevents efficient link utilisation at rates much above 64 kbit/s. For these reasons, frame relay communications protocol standards were proposed by CCITT in the late 1980s.

9.4   Frame relay service

Frame relay is being introduced by PTTs around the world as a service for the support of inter-LAN communications. The frame relay standard defines a simplified packet switching protocol in which error control and flow control overheads are minimised. Further, routing of data packets at network nodes is performed on link level addresses, thus minimising the protocol processing load. These features ensure a high data transfer rate with low transport delays on the transmission links together with high throughput at network nodes.

Initially, frame relay is being offered as a Permanent Virtual Connection (PVC) service across the wide area between corporate network sites. Service data rates of 64 kbit/s to 2 Mbit/s are available now with evolution to 34 Mbit/s (45 Mbit/s in North America) in the future.

Fig. 9.4 shows in simple terms the provision of a frame relay service. Frame relay is an ideal vehicle for point-to-multipoint communications for LAN interconnection in offering an HDLC-based variable length fast packet service. It is well supported by Customer Premises Equipment (CPE) vendors since it requires only an inexpensive software upgrade to existing CPE products (bridges/routers) for its implementation.

With the PVC service, the frame relay platforms or nodes in the public network contain routing tables which have been predefined by the network operator. A Data Link Connection Identifier (DLCI) field in the HDLC frame originated by a sender indicates the frame's destination. The frame relay nodes then simply look up the DLCI in their routing tables to determine the correct outgoing link on which to place the (unchanged) frame. A number of DLCIs can be made available to the corporate customer by agreement with the service provider. Each DLCI then enables the equivalent of a Permanent Virtual Connection (PVC) to be set up between the frame origination site and a given frame destination site. The transmission link between a corporate site and its parent frame relay platform is thus able to support several simultaneous communications sessions by means of frame interleaving, ensuring a high rate of bandwidth utilisation.

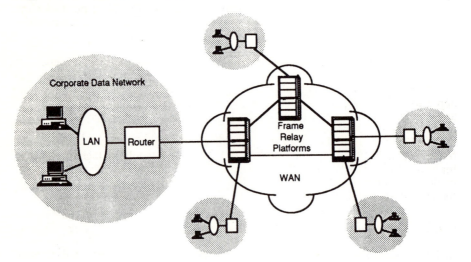

Fig. 9.4 *Frame relay service concept*

9.4.1 Frame format

The frame format used by frame relay is as shown in Fig. 9.5. It is based on LAPD as defined in CCITT Recommendation Q.922. Apart from the delimiting flags, the frame consists of three fields; an address/control field of 2 bytes, a variable length information field and a 2-byte frame check sequence (FCS).

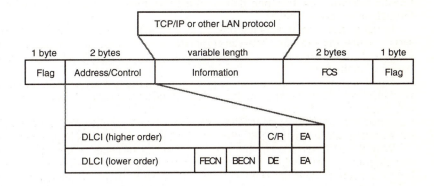

Fig. 9.5 *Frame relay frame format*

The address/control field is as follows :

- The 10 DLCI bits are used for addressing as indicated previously.

- The 2 EA bits are for extended address indications.

- The 1 bit Forward Explicit Congestion Notification (FECN), 1 bit Backward Explicit Congestion Notification (BECN) and 1 bit Discard Eligibility (DE) indicators are for flow control and congestion control purposes.

- The C/R bit indicates whether the frame is a command or a response frame.

The information field is variable in length with a maximum length defined for the application being supported, such that the maximum size frame produced by the application may be carried without segmentation. As shown above, the field is being used to carry a LAN protocol (such as TCP/IP).

The FCS field is used to check the frame for errors. When an error is detected at a frame relay node, the frame is discarded. Higher level protocols are then responsible for error recovery by ensuring that retransmission is initiated on an end-to-end basis by the application concerned. It should thus be clear that frame relay is only suitable for networks in which transmission links are likely to suffer low error rates.

9.4.2 Frame relay congestion control

In frame relay, there is no transmission 'window' system, relying on backward acknowledgement messages from the network to control data flow from the senders. Instead, a much simpler, more efficient system is implemented in which the network only sends messages to a sender in overload situations. This way, unnecessary protocol processing and reductions in achievable data throughput are avoided.

When a frame relay node determines the onset of congestion, it sends explicit congestion signals in both forward (towards destination) and backward (towards source) directions. Forward explicit congestion notification is provided by setting the FECN bit in the address field. Similarly, backward explicit congestion notification can be indicated by setting the BECN bit, but this is only useful if a timely frame is available in the reverse direction. Otherwise, link layer management informs the frame source of the congestion condition.

The purposes of explicit congestion notification are :

- to inform frame relay access nodes of the congestion so that they may take appropriate action; and/or

- to notify the frame source that the negotiated throughput has been exceeded.

The network then leaves it to the senders to adjust their data flows. If the senders take no action to alleviate the congestion, the network will discard frames.

Clearly, in a network that has been properly configured with sufficient resources, congestion should be a rare occurrence. However, in order to perform this network configuration, it is essential that the operator has knowledge of the demands likely to be placed on the network resources by customers. Towards this end, each customer is required to agree with the operator a Committed Information Rate (CIR) which determines the rate at which data can be sent on a given PVC from the customer's premises to his local frame relay access node. The total CIR for the link from customer to network is then the sum of the CIRs allocated to the individual PVCs. Since the data transmitted is bursty in nature, a stochastic process may be adopted. This enables a data source to transmit a Committed Burst Size (B_c) which uses a large proportion (or the whole) of the available link transmission capacity, but for a short enough time period (T_c) such that the agreed average CIR value is not exceeded. The CIR value is related to the Committed Burst Size (B_c) of data that may be transmitted per time period (T_c) by the formula $CIR = B_c/T_c$.

If a sender exceeds its allocated CIR, it should set the Discard Eligibility (DE) bit in those frames that are over the quota; alternatively, the access node will perform that function. A sender is allowed to exceed the allocated CIR value by a predetermined amount called the Excess Burst Size (B_e). If a sender exceeds its maximum quota ($B_c + B_e$) the excess frames are discarded by the network. Similarly, whenever congestion occurs, the B_e frames with DE bits set are always discarded by the network before the B_c frames without DE bit set.

In summary, the advantages of frame relay for the corporate customer are thus:

- it provides a point-to-multipoint service in which a single link from the customer's premises can provide simultaneous PVCs to a range of destination locations

- it can provide a degree of bandwidth flexibility in offering 64 kbit/s, n x 64 kbit/s and 2 Mbit/s in the short term with potential for up to 34 Mbit/s longer term

- it offers high throughput and efficiency since routing takes place at the link level within the network

- it can reduce costs for the corporate customer since inter-LAN communications can now be served by a single link into the public network rather than by multiple leased lines, whilst implementation costs in the CPE are negligible.

9.5 Switched Multimegabit Data Service

The Switched Multimegabit Data Service (SMDS) has been defined to meet the growing need for high performance LAN interconnection across metropolitan and wide areas. The service concept was originally introduced by Bellcore and is presently being deployed in many countries around the world, including the UK. It is the first widely available switched broadband public service and is capable of delivering bandwidth on demand for communications between a range of data applications.

SMDS is a packet-switched connectionless data service offering high speed internetworking between customer sites across public networks. It allows the exchange of variable length service data units (SDUs) or frames, with up to a maximum of 9188 octets of user data per frame. The performance, features and QOS provided by SMDS are similar to those available on LANs, including both high throughput and low delay.

For reasons of compatibility, SMDS has been defined as a connectionless service, since its principal aim is to support connectionless inter-LAN communications. A connectionless service has the merit of avoiding the delays and complexities normally associated with the call control signalling required with connection-oriented services. Instead, each SMDS frame carries its own address fields so that switching nodes in the SMDS network can route individual frames.

The connectionless switching functionality of SMDS is provided by one or more MAN Switching Systems (MSS) located in the public network. Fig. 9.6 illustrates a typical functional SMDS network architecture. It shows a number of customer sites, each supporting a corporate data network based on a LAN. A router directs appropriately addressed data frames from the LAN to one of the MSSs in the SMDS network. The MSS performs a routing function, based on the address fields carried in the header of each SMDS frame received, in order to place the frame on the correct outgoing path towards its destination.

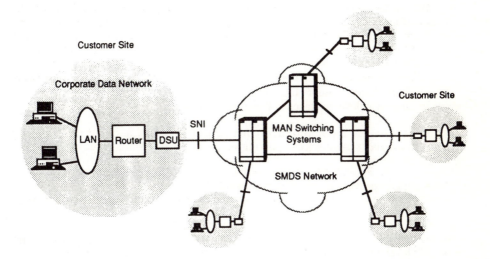

Fig. 9.6 *SMDS network architecture*

Access from customer sites to the SMDS network takes place across the Subscriber Network Interface (SNI). The protocol used at the SNI is the SMDS Interface Protocol (SIP), based on the IEEE 802.6 DQDB access protocol. As required by DQDB, communications across the SNI take place over a pair of unidirectional buses, one carrying transmissions from customer premises to MSS and the other from MSS to customer premises. DQDB is based on a cell-based protocol. Since the costs associated with upgrading multiple types of CPE (such as a router) to convert from the frame-based LAN protocol to the cell-based SIP would be unattractive, because of the relatively small quantities of each type, specialised Data Service Units (DSUs) which perform this function have become available, as shown in the figure.

9.5.1 Protocol architecture

A network providing SMDS is intended to assume the role of a subnetwork in the context of a customer's communications architecture. This means that the SMDS network's SIP protocol carries the customer's internetworking protocols transparently, just as a MAC protocol carries an internet protocol across a LAN. These principles are illustrated in Fig. 9.7, which shows a protocol architecture based on the SMDS network described by Fig. 9.6.

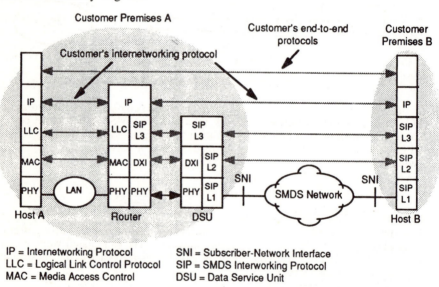

Fig.9.7 *Customer's internetworking protocol architecture*

In Fig. 9.7, Host A on customer premises A carries a typical LAN-oriented protocol stack by which means it is able to communicate with the router. The router converts the data frames from the LAN into SIP level 3 PDUs complete with SMDS E.164 addresses, translated from the higher layer IP addresses. It then encapsulates the SIP level 3 PDUs into the HDLC-based Digital Exchange Interface (DXI) protocol for transfer to the DSU. The DSU provides an SMDS protocol conversion capability that supports the SIP protocol stack on the network side and DXI on the customer side. At the DSU, SIP level 3 PDUs from the router are converted, by a segmentation process, into SIP level 2 PDUs, which are fixed length 53 octet cells. These are transferred to the underlying SIP level 1 physical layer for transmission across the SNI to the MSS. The MSS switches the cells to their destination on customer premises B, where receiving host B, capable of operating the full SIP stack, reassembles them and converts them back into their original higher layer format for transfer to the receiving application.

Because the SIP is based on the IEEE 802.6 DQDB protocol, CPE that operates the latter can access SMDS without the need for the SIP conversion functionality exemplified by the DSU. Under these circumstances, two possible access configurations are possible as shown in Fig. 9.8.

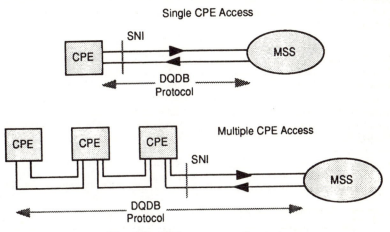

Fig. 9.8 *CPE access configurations*

The simpler configuration connects a single DQDB conformant CPE (workstation, PC, etc.) over the subscriber network interface to a MSS. The more complex configuration shows multiple CPE operating as a DQDB system, not only for inter-CPE communications but also as a means of sharing an SNI for access to the MAN switching system.

9.5.2 Protocol format

The SIP is a 3-level protocol stack - the term 'level' is used to distinguish the stack from the OSI layered model. The three protocol levels include addressing, framing, error detection and physical transmission functions. The functionality encompassed by levels 3 and 2 of the SIP is shown in the Fig. 9.9 by reference to the encapsulation of information at each level.

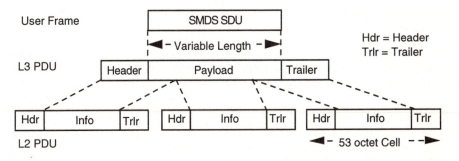

Fig. 9.9 *SIP encapsulation mechanism*

Above SIP level 3 is the HDLC frame (SMDS SDU) as it arrives at, say, a router from a LAN application. This is encapsulated by the addition of a header and a trailer field in the router to provide the level 3 PDU. The header contains the E.164 source and destination addresses for the PDU; the trailer contains a means to enable lost level 2 PDUs to be detected.

The format of a SIP level 3 PDU and a brief explanation of the fields carried is given in Fig. 9.10.

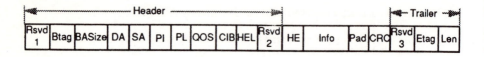

Fig. 9.10 *SMDS interface protocol level 3 PDU*

- Rsvd 1 : the 1-octet reserved field that is currently unused.

- Btag : the 1-octet beginning tag field carries a binary value of from 0 to 255, used to form an association between the first and last cells of an L3_PDU. For this reason, the same binary value is inserted into the Etag of the PDU trailer.

- BASize : the 2-octet Buffer Allocation Size field gives the length in octets of the PDU, from the start of the DA field up to and including the pad field and (if present) the CRC field.

- DA : the 8-octet destination address field carries the destination address for the L3_PDU.

- SA : the 8-octet source address field carries the source address of the L3_PDU.

- PI : the 6-bit protocol identifier field is used to indicate the service user to which the Info field is to be sent at the PDU destination.

- PL : the 2-bit pad length field is used to indicate the length in octets of the pad field.

- QOS : the 3-bit quality of service field indicates the requested quality of service for the L3_PDU.

- CIB : the CRC indicator bit conveys the presence or absence of the CRC field in the L3_PDU.

- HEL : the 3-bit header extension length field gives the length of the header extension field of the PDU. This length in octets is obtained by multiplying the value of HEL by four.

- Rsvd 2 : a 2-octet reserved field that is currently unused.

- HE : the header extension field is of variable length from 0 to 20 octets, included to convey additional protocol control information that may be standardised in the future.

- Info : the variable length information field has a maximum length of 9188 octets and conveys user information.

- Pad : the variable length padding field of from 0 to 3 octets ensures that the total length of info plus pad fields is an integral multiple of four octets.

- CRC : the optional 4-octet Cyclic Redundancy Check field is used for error detection and is calculated over all PDU fields from the start of the DA field to the end of the pad field.

- Rsvd 3 : a 1-octet reserved field that is currently unused.

- Etag : the 1-octet end tag carries the same binary value as the Btag field.

- Len : the 2-octet Length field indicates the length in octets of the PDU from the start of the DA field to the end of the CRC field (if present). It carries the same value as the BASize field in the PDU header.

At SIP level 2, segmentation and reassembly is performed, typically by a DSU, on the variable length L3 PDUs. The 53 octet fixed length cells used at this level carry their own header and trailer fields, the latter of which includes functionality for bit error detection. The headers also carry information which identifies the L2 PDUs as the Beginning of Message (BOM), Continuation of Message (COM) or the End of Message (EOM) cell, or alternatively as a Single Segment Message (SSM).

The format of a SIP level 2 PDU and brief explanation of its fields is given in Fig. 9.11. Note that the 5-octet cell header is formatted identically with the cell headers for DQDB cells and that some of the fields will not be used by the SMDS.

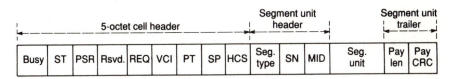

Fig. 9.11 *SMDS interface protocol level 2 PDU*

- Busy : the single bit busy field indicates whether the cell contains information or is empty.

- ST : the single bit Slot Type field is used by the DQDB protocol to indicate the type of information being carried by the cell.

- PSR : the single bit Previous Slot Reset field is used by the DQDB protocol to indicate that the information carried in the previous cell can be cleared.

- Rsvd : a 2-bit reserved field that is currently unused.

- Req : the 3-bit request field is used by the DQDB protocol to indicate the priority level at which the payload is to be buffered in the distributed queue.

- VCI : the 20-bit Virtual Channel Identifier field is used by the DQDB protocol to indicate the virtual channel to which the segmentation unit belongs.

- PT : the 2-bit Payload Type field is used by the DQDB protocol to indicate the type of information being carried in the cell payload.

- SP : the 2-bit Segment Priority field is reserved for future use.

- HCS : the 8-bit Header Check Sequence field provides for the detection of errors and the correction of single bit errors in the 5-octet cell header.

- Seg Type : the 2-bit segment type field indicates how the cell should be processed by the receiving entity (i.e. BOM, COM, EOM or SSM segment).

- SN : the 4-bit Sequence Number field is used to verify that all of the cells (level 2 PDUs) of the level 3 PDU have been received and concatenated in the correct order.

- MID : the 10-bit Message Identifier field is used to associate all the cells of a given level 3 PDU, which are accordingly labelled with the same MID number. For SSM cells, the MID value is set to all zeros.

- Seg Unit : the segmentation unit comprises the payload of the SIP level 2 PDUs, carrying a segment of level 3 PDUs.

- Pay Len : the 6-bit payload length field indicates the number of octets of the Level 3 PDU carried by the segmentation unit (always a multiple of four octets between 4 and 44 inclusive). The value of payload length is set to 44 for BOM and COM messages; for EOM and SSM, the value is as appropriate.

- Pay CRC : the 10-bit payload Cyclic Redundancy Check field is used for detection of errors and correction of single bit errors in them level 2 PDU. The fields covered by the CRC include the segmentation unit header, segmentation unit and payload length field of the segmentation unit trailer.

Below level 2, SIP level 1 provides the physical interface to the SMDS network's digital transmission facilities. To accomplish this, SIP level 1 is divided into two parts, performing Physical Layer Convergence Procedure (PLCP) functions and transmission system functions, respectively.

The transmission system defines the characteristics of the SMDS network's digital transmission facilities, i.e. the links between a customer site and an MSS. Multiple types of transmission systems are possible, reflecting the range of access link options available to a customer, e.g. 2 Mbit/s, 34 Mbit/s, etc. SMDS uses transmission systems consistent with the standard carrier network hierarchy; such systems use a framing technique for the carriage of information. Frames occur in a regular, periodic manner and are delimited by special framing octets. Other overhead octets, plus SIP level 2 PDU octets and SIP level 1 control information octets, are interleaved into the frames at the transmitter in the same relative positions for ease of recovery at the receiver.

The PLCP adapts the capabilities of the transmission system to the requirements of the SIP level 2. It defines the manner in which the SIP level 2 PDUs and the SIP level 1 control information are mapped into frames for transfer by the underlying transmission system. Thus, a unique PLCP is required for use with each different individual transmission system.

Note that SIP level 1 control information consists of two octets. These are used to provide CPE with information on the network configuration and on bus identification (on which bus to transmit and on which to receive data), as well as for MID allocation purposes.

As shown in Fig. 9.10, all SMDS level 3 PDUs carry a Source Address (SA) and a Destination Address (DA) of 8 octets each, formatted according to the E.164 addressing standard. If the level 3 PDU is of length not greater than 44 octets, it fits into a single cell (level 2 PDU) with the ST field set to indicate a Single Segment Message (SSM). Otherwise, following segmentation of the level 3 PDU, the ST field of the first cell is set to indicate Beginning of Message (BOM) with the ST fields in subsequent cells set to Continuation of Message (COM) and the ST field in the final cell set to End of Message (EOM). Since the DA of the level 3 PDU appears only in the BOM segment, some means is required to enable all of the segments of the level 3 PDU to be associated for reassembly at the receiver. For this purpose, every cell of a given level 3 PDU carries an identical Message Identifier (MID). The MID associated with any L3 PDU is always unique at any given instant on any given access path to a customer's premises (i.e. at any given SNI). This enables multiplexing of cells belonging to different L3 PDUs across an SNI between the customer's premises and an MSS.

Source Addresses (SA) carried by the level 3 PDUs are validated by SMDS to ensure that they have been legitimately assigned by the originating CPE. This is to prevent either error or fraud from causing call charges to be associated with the wrong customer.

Another addressing feature supported by SMDS is group addressing, where a single address is used to identify a set of destination addresses. A group addressed level 3 PDU results in multiple copies being made by the SMDS, which then forwards one to each of the individual addresses represented by the group address. Address screening is also provided; this feature allows customers to specify either source addresses from which level 3 PDUs may (or may not) be received or destination addresses to which level 3 PDUs may (or may not) be sent.

In operation, cells transmitted across an SNI from a customer site are typically routed through the public transport and switching network before they arrive at an MSS. On receipt of the first cell of a new incoming L3 PDU, an MSS notes the destination address being carried and, from this, determines the most appropriate outgoing path towards the destination. The MSS also notes the MID carried by this first cell; the MID/DA association will be used by the MSS to switch subsequent cells belonging to the same L3 PDU towards the same destination. However, there is no guarantee that all cells of the same L3 PDU will reach their destination via the same route. The connectionless protocol enables a routing decision to be based on prevailing network conditions at the time of arrival of each cell, to optimise network usage and minimise delays and congestion.

The access path from an SMDS customer site to the network is dedicated to that customer, in order to guarantee privacy and security. Data are exchanged across the SNI at either E1 (2 Mbit/s) or E3 (34 Mbit/s) rates (in Europe), with prospects for higher rates of up to 155 Mbit/s in the future. For E1 transport, a single access class is available, whereas for E3 up to 4 access classes are supported. The access class determines the average rate of data transfer across the SNI, called the Sustained Information Rate (SIR). Access classes allow the data rates into and out of the network to be regulated for consistency with the capabilities of the CPE. They allow the network to allocate resources efficiently and protect CPE from being overwhelmed by data. The four SIRs available on E3 are 4 Mbit/s, 10 Mbit/s, 16 Mbit/s and 25 Mbit/s. The SMDS enforces the SIR to ensure that the data rate associated with the access class at a particular CPE is not exceeded. Violation of the SIR results in the level 3 PDU being discarded.

In summary, some of the advantages of SMDS to a corporate customer are :

- it is the first widely available public wide area broadband data service

- it is a connectionless service that allows multicast operation, which are ideal characteristics for the support of inter-LAN communications

- it is a switched service allowing point-to-multipoint operation without the need to provision PVCs or to predict in advance the network addresses with which communications may be required

- it uses the standard E.164 addressing for unrestricted global communications

- it is a cell-based service that anticipates ATM

- it ensures secure communications by virtue of its address validation and screening mechanisms.

9.6   Broadband ISDN support for data services

Frame relay applications and services as well as connectionless data applications and services can be expected to be required for the foreseeable future. This means that both frame relay and connectionless data protocols and bearer services will need to be supported by B-ISDN and its Asynchronous Transfer Mode (ATM) transport and switching services. There are many potential scenarios which illustrate different aspects of the support provided by a B-ISDN to data services; a number of such scenarios are described in the following.

9.6.1 Broadband ISDN support for frame relay services

9.6.1.1 Frame Relay Bearer Service emulation
The simplest frame relay-based scenario concerns a single ATM connection set up over the B-ISDN between two customer premises. The B-ISDN provides a basic ATM transport for the Frame Relay Bearer Service (FRBS) emulated in the two FR/B-ISDN terminals involved in the connection. The FRBS emulation capability provides for the unacknowledged transfer of data frames between user-network interfaces (UNIs) at the S_B or T_B reference points. Both the network architecture and the protocol architecture associated with this scenario are shown in Fig. 9.12. Two B-ISDN terminals using the FRBS are connected over the B-ISDN by means of appropriate ATM switching elements. These may be ATM cross-connect equipment for handling Permanent Virtual Connections (PVCs) or ATM switches for handling Switched Virtual Connections (SVCs), or possibly a mixture of the two.

The protocol stacks exemplified in the FR/B-ISDN terminals consist of upper layers (whose details are of no concern in the present context) being served by an ATM Adaptation Layer Type 5 (AAL5), then an ATM layer and, finally, the physical layer. The AAL5 is subdivided into a Frame Relay Service Specific Convergence Sublayer (FR-SSCS) which is used to emulate the FRBS in B-ISDN, a Common Part Convergence Sublayer (CPCS) and a Segmentation and Reassembly Sublayer (SAR).

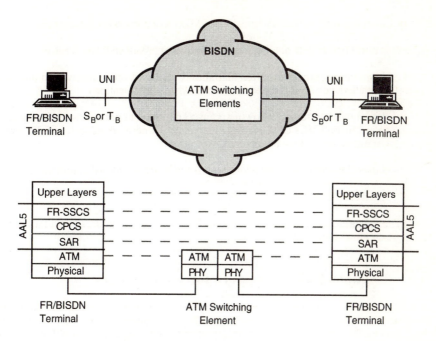

Fig. 9.12 *FRBS emulation in B-ISDN*

At the Service Access Point (SAP) between the upper layers and the FR-SSCS, FR-SSCS-SDUs are exchanged. On receipt of an SDU from the upper layers, the FR-SSCS constructs a PDU for transfer to the peer FR-SSCS entity in the distant terminal. In the other direction, on receipt of a PDU by the FR-SSCS from its distant peer entity, an FR-SSCS-SDU is constructed for transfer to the upper layers. The capability to carry multiple FR-SSCS connections over a single CPCS connection is provided by multiplexing, using the DLCI field to differentiate the connections.

The format of the FR-SSCS-PDU is as shown in Fig. 9.13. The destination address carried by the FR-SSCS-PDU (i.e. the DLCI) is provided by layer management. The address is determined either by call control signalling during connection establishment in the case of SVCs, or at subscription time in the case of PVCs.

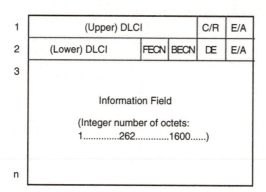

Fig. 9.13 *Format of FR-SSCS-PDU*

Note that the FR-SSCS-PDU format is identical with the frame format described previously in Section 9.4 but without the delimiting flags and the FCS field. The above format is mandatory; other formats with three or four octet headers are optional.

A similar exchange of primitives (SDUs) as described earlier occurs at the interface between the FR-SSCS and the CPCS. The FR-SSCS-PDU becomes the CPCS-SDU which is encapsulated into a CPCS-PDU for transfer to the distant CPCS entity. Encapsulation in the case of the AAL5 CPCS-PDU involves the addition of an 8-octet trailer to the CPCS-SDU. The encapsulation process used in the AAL5 CPCS and the subsequent CPCS-PDU format are as shown in Fig. 9.14.

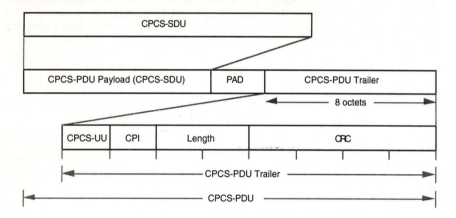

Fig. 9.14 *CPCS-PDU format for AAL Type 5*

The fields shown in the CPCS-PDU are as follows:

- CPCS-PDU payload : used to carry the CPCS-SDU. The field is octet aligned and is from 1 to 65,535 octets in length.

- PAD : this is a padding field of from 0 to 47 octets in length, used as filler to provide 48 octet alignment for the CPCS-PDU.

- CPCS-UU : the CPCS User-to-User indication field of one octet is used to transfer transparently CPCS user-to-user information.

- CPI : the Common Part Indicator field of one octet aligns the CPCS-PDU trailer to 64 bits.

- Length : the length field of two octets carries the length of the CPCS-PDU payload field.

- CRC : the Cyclic Redundancy Check field of 4 octets is used to detect bit errors in the CPCS-PDU.

At the interface between the CPCS and the SAR sublayer, SAR-SDUs are exchanged. The SAR sublayer accepts variable length SAR-SDUs, comprising an integral multiple of 48 octets, from the CPCS. From these SDUs, the SAR sublayer performs a segmentation function to construct 48 octet SAR-PDUs containing SAR-SDU data. The SAR-PDUs become the ATM-SDUs which are transferred to the underlying ATM layer. Here, a 5 octet header is appended to form ATM-PDUs (ATM cells) of 53 octets which are passed down to the physical layer for transmission onto the medium. Likewise, in the other direction of transfer, ATM-SDUs are recovered in the ATM layer from the incoming ATM cells received from the physical layer. The ATM-SDUs (SAR-PDUs) are transferred to the SAR sublayer and re-assembled into SAR-SDUs for transfer to the CPCS.

The coding and structure of the SAR-PDU and the ATM-PDU are shown in Fig. 9.15. The SAR sublayer provides the ATM-layer-user to ATM-layer-user (AUU) parameter to indicate whether a SAR-PDU carries the end or the beginning/continuation of the SAR-SDU. An AUU value of '1' indicates the SAR-PDU carries the end of the SAR-SDU whilst a value of '0' indicates either beginning or continuation of the SAR-SDU. Note that the AUU parameter is conveyed in the header of the ATM-PDU (ATM cell) as part of the Payload Type (PT) field, rather than as part of the SAR-PDU.

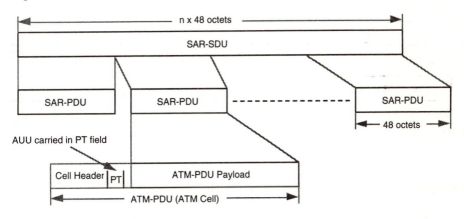

Fig. 9.15 *SAR-PDU and ATM-PDU formats*

As described above, FR-SSCS-SDUs are processed down through the protocol stack in one of the terminals and transmitted as a bitstream across the physical medium to the ATM switching elements. Here, the bitstream is processed upwards through the protocol stack, but only up to the ATM layer, where the ATM PDUs are recovered. This enables a routing and switching function to be performed, based on the VPI/VCI value carried in the ATM-PDU header. The ATM-PDUs are then reconstituted with new VPI/VCI values (if necessary) so that the routing and switching process may be continued in other ATM switching elements on the path to the destination terminal. Finally, at the destination terminal, the original FR-SSCS-SDU is recovered and passed up to the higher layers.

9.6.1.2 Integral Frame Relay Bearer Service
In this scenario, the FRBS is provided within the B-ISDN by means of one or more Frame Relay Service Functions (FRSF). The FRSF handles frame relay protocols and routes and relays data according to routing information provided either during frame relay connection establishment in the case of SVCs or at service subscription time in the case of PVCs. The network architecture and the protocol architecture for this scenario are illustrated in Fig. 9.16.

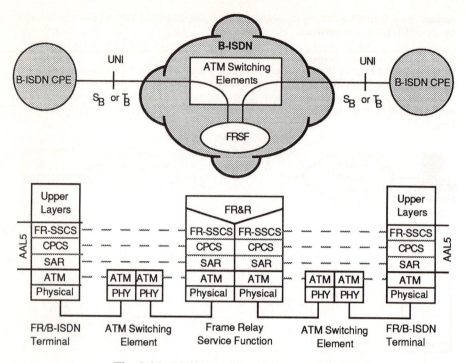

Fig. 9.16 *FRBS provision integral with B-ISDN*

Two B-ISDN CPEs using the FRBS are connected across the B-ISDN via ATM switching elements. The frame relay protocols are carried over ATM connections and through the ATM switching elements to one or more of the FRSFs provided within the B-ISDN (for convenience, only one FRSF is shown). The FRSF handles the frame relay protocols between the two CPEs. It performs a routing and relay function, based on the DLCIs carried within the frame headers, to place incoming frames received from one CPE on the correct outgoing ATM connection towards the destination CPE.

The protocol stacks implemented in the two communicating end terminals within the CPEs are identical with those described in the previous scenario. The processing of information through the layers and the formats of the information within each layer or sublayer are also identical, as are the switching and routing of the information through the ATM switching elements to the FRSFs.

At the FRSF, the received bitstream is processed up through the protocol stack to recover the FR-SSCS-PDU. Within the Frame Routing & Relay (FR&R) sublayer, the DLCI in the frame header of the FR-SSCS-PDU is converted to a new DLCI value, for insertion into the frame as it is relayed on its new path from the FRSF towards the destination terminal. At the ATM layer on the outgoing side of the FRSF, the VPI/VCI value for the ATM-PDUs is provided by layer management, from information gained either by call control signalling during connection establishment in the case of SVCs, or at subscription time in the case of PVCs. The VPI/VCI ensures that the frames of data are placed on the desired ATM connection towards the destination CPE.

In this scenario, statistical multiplexing is performed at the frame level. In other words, a single ATM connection may be used to carry multiple frame relay connections, differentiated by means of their DLCIs and handled for routing purposes by the FRSF(s) in the network.

9.6.1.3 Interworking between FRBS and B-ISDN

In this third scenario, interworking between frame relay and B-ISDN networks is described. An Interworking Function (IWF) is used at the boundary of the two networks to provide the protocol conversion capability needed for interworking. The network architecture and the protocol architecture for this scenario are illustrated in Fig. 9.17.

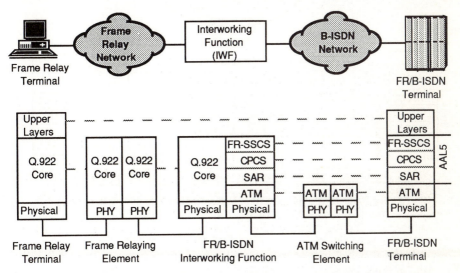

Fig. 9.17 *Network interworking between FRBS and B-ISDN*

The scenario shows a B-ISDN terminal supporting the FRBS connected across the B-ISDN via one or more ATM switching elements to an IWF. From the IWF, a connection exists over a frame relay network via one or more frame relaying elements to a FR terminal. The IWF transfers unchanged the information fields of the PDUs between the FR-SSCS and the Q.922 core layer to achieve interworking. The protocol control information derived from the headers of the two interworked protocols (FR-SSCS and Q.922 Core) is exchanged by means of parameters conveyed in primitives. These parameters are processed by the IWF to create the correct headers for the PDUs of each of the interworked protocols. The use of the B-ISDN by the frame relaying network is invisible to the end users, which allows the upper layer protocols to be exchanged intact.

The protocol stacks implemented in the FR/B-ISDN terminal and on the B-ISDN side of the IWF are identical with those described in the previous scenarios. The processing of information through the layers and the formats of the information within each layer or sublayer are also identical, as are the switching and routing of the information through the ATM switching elements.

The frame relay terminal implements a suite of upper layer protocols which are served by the Q.922 core layer, running over a physical layer. The frame relaying elements in the FR network and the FR side of the IWF also implement Q.922 core and PHY (physical) layers. The format of the Q.922 core-PDUs exchanged between the FR terminal and the IWF are as shown in Fig. 9.18.

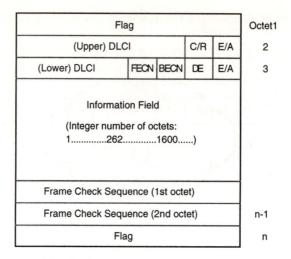

Fig. 9.18 *Format of Q.922 core-PDU*

Note that the Q.922 core-PDU format is identical with the frame format described previously in Section 9.4.

Q.922 core-PDUs originated by the FR terminal are routed by the frame relaying elements within the frame relay network towards the IWF. Here, the flags and FCS are stripped off and the protocol control information contained in octets 2 and 3 is modified as appropriate to form the FR-SSCS-PDUs as already described. The FR-SSCS-PDUs are then transferred across the B-ISDN network to the FR/B-ISDN terminal where the upper layer protocols are recovered. An inverse series of operations is performed in the other direction of transmission.

9.6.2 Broadband ISDN support for broadband connectionless data service

9.6.2.1 Provision of broadband connectionless data bearer service by a single connectionless server

In the simplest scenario, the Broadband Connectionless Data Bearer Service (BCDBS) is provided within the B-ISDN by means of a single Connectionless Service Function (CLSF), which is implemented within a B-ISDN element called a Connectionless Server (CLS). The CLSF handles the connectionless data and routes and relays the data according to addressing information carried by the connectionless data PDUs. This scenario is similar to that described for SMDS which was provided within a MAN by means of an MSS. The difference is that the BCDBS uses B-ISDN protocols as will be described, instead of the SIP protocols of SMDS. The network architecture and protocol architecture for this scenario are illustrated in Fig. 9.19.

Two B-ISDN CPEs using the BCDBS are connected across the B-ISDN via ATM switching elements. Information is transferred in a connectionless manner between the CPEs by means of the Connectionless Network Access Protocol (CLNAP). The CLNAP is used at the ATM User Network Interface (UNI) and is carried over ATM connections and through the ATM switching elements to the CLSF provided within the B-ISDN. Note that the simplest possible arrangement for this scenario would be with direct connections between the CPEs and the CLSF. However, the more general arrangement with interposed ATM switching functionality has been chosen for illustration.

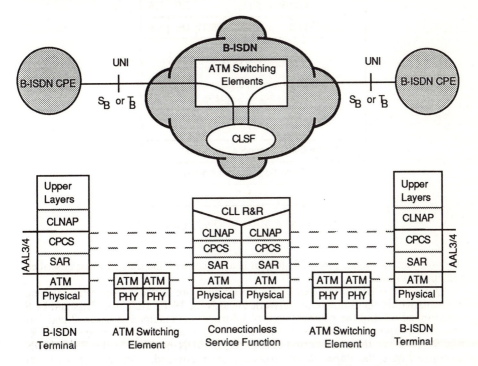

Fig. 9.19 *BCDBS provision by single CLSF within B-ISDN*

The CLSF terminates the B-ISDN connectionless protocol (CLNAP) and adapts this protocol to the connection-oriented ATM layer protocol. The CLSF also performs a routing and relay function, based on the E.164 addresses carried within the headers of the connectionless data PDUs, to place incoming data received from one CPE on the correct outgoing ATM connection towards the destination CPE.

The protocol stacks implemented in the B-ISDN CPEs consist of upper layers being served by a Connectionless Layer (CLL) in which the CLNAP resides. The CLL provides for the transparent transfer of variable size CLL SDUs from a source to one or more destination CLL user(s) such that lost or corrupted data units are not retransmitted. In turn, the CLL is supported by an ATM Adaptation Layer Type 3/4 (AAL3/4), then an ATM layer and finally, the physical layer. The AAL3/4 is subdivided into a Common Part Convergence Sublayer (CPCS) and a Segmentation and Reassembly Sublayer (SAR). The service specific convergence sublayer is null in this instance of AAL3/4 application.

At the SAP between the upper layers and the CLL, CLNAP-SDUs are exchanged. On receipt of an SDU from the upper layers, the CLL constructs a CLNAP-PDU for transfer to the peer CLL entity in the distant CPE. In the opposite direction, on receipt of a CLNAP-PDU from the distant CPE, the CLL constructs an SDU for transfer to the upper layers. The format of the CLNAP-PDU is as shown in Fig. 9.20.

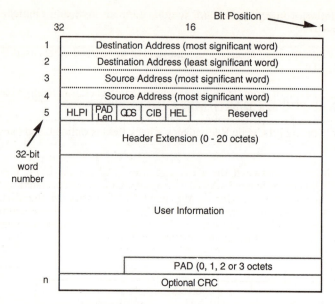

Fig. 9.20 *Format of CLNAP-PDU*

The fields shown in the CLNAP-PDU are as follows:

*   Destination address : an 8 octet field containing a 4 bit 'address type' subfield and a 60 bit 'address' subfield; the latter is structured according to the ITU-T Recommendation E.164.

*   Source address : an 8 octet field structured as indicated above for the destination address.

*   HLPI : the Higher Layer Protocol Identifier is a 6 bit field used for identification of the CLNAP user or OAM entity with which the CLL-SDU is to be associated at the destination node.

*   PAD Len : the 2 bit PAD length field indicates the length of the PAD field in octets. The length of the PAD field plus the length of the user information field is required to be an integral multiple of four octets.

*   QOS : the 4 bit Quality of Service field is used to indicate the QOS requested for the CLNAP-PDU.

*   CIB : the single CRC Indication Bit field indicates the presence (= 1) or absence (= 0) of a 32 bit CRC field.

*   HEL : the Header Extension Length field of 3 bits carries a value of from 0 to 5 and indicates the number of 32 bit words in the header extension field.

*   Reserved : a 16 bit field reserved for future use.

*   Header extension : a variable length field in the range from 0 to 20 octets carrying structured information elements. An element comprises three

subfields, conveying element length, element type and element payload, respectively.

- User information : a variable length field of up to 9188 octets used to carry the CLL-SDU.

- PAD : a field of from 0 to 3 octets whose length is chosen to align the resulting CLNAP-PDU to a 32 bit boundary.

- CRC : the 32 bit Cyclic Redundancy Check field is optional and carried end to end.

At the interface between the CLL and the CPCS of AAL3/4, CPCS-SDUs (CLNAP-PDUs) are exchanged. An SDU from the CLL is encapsulated by the CPCS into a CPCS-PDU for transfer to the peer CPCS entity in the distant CPE. Encapsulation in the case of the AAL3/4 CPCS-PDU consists of the addition of a 4 octet header and a four octet trailer to the CPCS-SDU. The encapsulation process and the resulting CPCS-PDU format are as shown in Fig. 9.21.

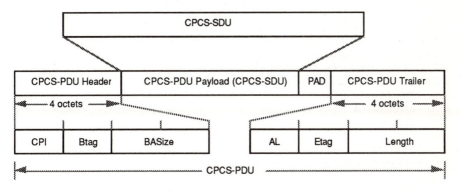

Fig. 9.21 *CPCS-PDU format for AAL Type 3/4*

The fields carried by the CPCS-PDU are as follows:

- CPI : the Common Part Indicator field of 1 octet is used to interpret subsequent fields in the CPCS-PDU header and trailer.

- Btag : the 1 octet beginning tag field allows the CPCS-PDU header and trailer to be correctly associated at the receiver. It carries the same value as the Etag field in the trailer and the value is incremented for each successive CPCS-PDU.

- BASize : the 2 octet Buffer Allocation Size field indicates to the receiver the maximum buffer requirements to receive the complete CPCS-SDU.

- CPCS-PDU payload : this is a variable length field which carries the CPCS-user information, i.e. the CPCS-SDU.

- PAD : the padding field contains from 0 to 3 unused octets which align the CPCS-PDU payload to an integral number multiple of four octets.

- AL : the Alignment Field is a single unused octet which achieves 32 bit alignment for the CPCS-PDU trailer.

- Etag : the end tag field carries the same value as the Btag field for a given CPCS-PDU in order to permit the association of the CPCS-PDU header with the CPCS-PDU trailer.

- Length : the length field indicates the length of the CPCS-PDU payload and enables the receiver to detect loss or gain of information.

Below the AAL3/4 CPCS resides the SAR sublayer and SAR-SDUs (CPCS-PDUs) are exchanged at this interface. The SAR sublayer accepts variable length SAR-SDUs from the CPCS on which it performs a segmentation function. The resulting fixed length, 48 octet SAR-PDUs contain 44 octets of SAR-SDU information (segmentation unit) plus a 2 octet header and a 2 octet trailer. The SAR-PDUs become the ATM-SDUs which are transferred to the underlying ATM layer. Here, a 5 octet header is appended to form ATM-PDUs (ATM cells) of 53 octets which are passed down to the physical layer for transmission onto the medium. Likewise, in the other direction of transfer, ATM-SDUs are recovered in the ATM layer from the incoming ATM cells received from the physical layer. The ATM-SDUs (SAR-PDUs) are transferred to the SAR sublayer and re-assembled into SAR-SDUs for transfer to the CPCS.

The coding and structure of the AAL3/4 SAR-PDU and the ATM-PDU are shown in Fig. 9.22. Note that the segmentation of the SAR-SDU may result in a final segmentation unit of an integer multiple of 4 octets that is less than 44 octets in length. This will result in a SAR-PDU carrying a payload field in which a multiple number of 4 octets are unused.

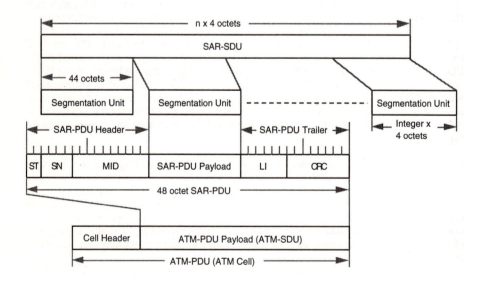

Fig. 9.22 *AAL3/4 SAR-PDU and ATM-PDU formats*

The fields carried by the SAR-PDU are as follows:

- ST : the 2 bit Segment Type field indicates whether the SAR-PDU carries a Beginning of Message (BOM), Continuation of Message (COM), End of Message (EOM) or Single Segment Message (SSM).

- SN : the 4 bit Sequence Number field allows the SAR-PDUs belonging to a given CPCS-PDU to be numbered modulo 16. Each SAR-PDU has its SN incremented by one relative to the previous SAR-PDU, which allows the receiver to detect loss or gain of PDUs.

- MID : the 10 bit Multiplexing Identification field is used to identify SAR-PDUs belonging to a given SAR-SDU. Its use permits multiplexing and subsequent recovery of different SAR-SDUs over the same ATM connection.

- SAR-PDU payload : the 44 octet payload field carries the SAR-SDU information (segmentation unit) in left-justified manner. For an SSM or EOM segment of less than 44 octets, this results in unused octets which are set to '0' and ignored at the receiver.

- LI : the 6 bit Length Indication field indicates the number of SAR-SDU octets carried in the payload field.

- CRC : the 10 bit Cyclic Redundancy Check field is used to detect bit errors in the SAR-PDU.

Following from the above, it should be clear that CLNAP-SDUs are processed down through the protocol stack in an originating terminal and transmitted as a bitstream across the physical medium to the ATM switching elements. Here, the bitstream is processed up through the protocol stack as far as the ATM layer, where the ATM-PDUs are recovered. This enables a routing and switching function to be performed, based on the VPI/VCI value carried in the ATM-PDU header, to place the ATM-PDUs on the correct path towards the CLSF. At the CLSF, the received bitstream is again processed up through the protocol stack to recover the CLNAP-PDU. The CLL R&R function then performs a routing operation based on the E.164 destination address carried by the PDU. This enables ATM-PDUs to be reconstituted with new VPI/VCI values so that the routing and switching process may be continued in other ATM switching elements on the path to the destination terminal. Finally, at the destination terminal, the original CLNAP-SDU is recovered and passed up to the higher layers.

9.6.2.2 *Provision of broadband connectionless data bearer service by multiple connectionless servers*

In this scenario, the BCDBS is provided within the B-ISDN by means of more than one connectionless service function. This involves the transparent transfer of connectionless service data units between CLSs using the ATM switching capabilities of the B-ISDN. The network and protocol architectures for this scenario are illustrated in Fig. 9.22.

As before, two B-ISDN CPEs using the BCDBS are connected across the B-ISDN via ATM switching elements, using the Connectionless Network Access Protocol (CLNAP) for information transfer. From one of the CPEs, data is carried by means of the CLNAP over ATM connections and through the ATM switching elements to the first CLSF provided within the B-ISDN. Here, a CLL R&R function performs a process to transform the incoming CLNAP-PDUs into CLNIP-PDUs, which must be used for connectionless data transfer at the ATM Network Node Interface (NNI) that exists between the two CLSFs. The CLL R&R function at the second CLSF then processes the incoming CLNIP-PDU to recover the original CLNAP-PDU for transfer over the network to the destination CPE. Note that, although only two CLSFs have been shown for simplicity, the transfer of CLNIP-PDUs between multiple transit CLSFs is covered below.

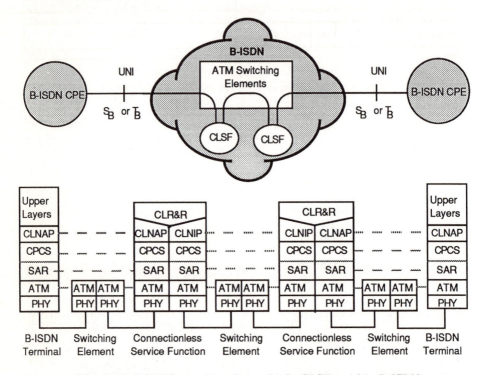

Fig. 9.23 *BCDBS provision by multiple CLSFs within B-ISDN*

The protocol stacks implemented in the two communicating end terminals within the CPEs are identical with those described in the previous scenario. The processing of information through the layers and the formats of the information within each layer or sublayer are also identical, as are the switching and routing of the information through the ATM switching elements to the CLSF.

At the first CLSF, the received data is processed up through the protocol stack to recover the CLNAP-PDU. The CLL R&R function then operates on the CLNAP-PDU according to whether the CLSF functions are located in separate operators' domains or within a single operator's domain. The following rules apply in each case:

1. For NNIs located in different operators' domains, encapsulation of the CLNAP-PDU is the requirement. This is performed in the CLL R&R function, which adds a 4 octet alignment header to the CLNAP-PDU to form the CLNIP-SDU. The CLNIP entity within the CLL then accepts this SDU to which it adds a further 40 octet header to form the CLNIP-PDU. The CLNIP-PDU in turn becomes the CPCS-SDU, which is processed as described previously. The encapsulation mechanism and resulting CLNIP-PDU format are as shown in Fig.9.24.

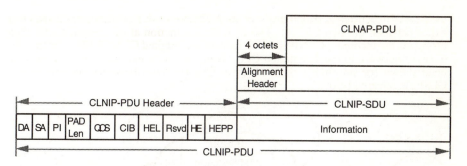

Fig. 9.24 *Encapsulation of CLNAP-PDU within a CLNIP-PDU*

The fields of the CLNIP-PDU carry identical meanings and lengths as the corresponding CLNAP-PDU fields, except for:

- PI : the 6 bit Protocol Identifier field indicates whether the PDU is encapsulating or not. For non-encapsulation, it is identical with the CLNAP-PDU HLPI field.

- HEPP : the Header Extension Post Pad field has a length in octets of 20 minus the length of the HE field, in order to ensure a CLNIP-PDU header length of 40 octets. In the case of a non-encapsulating CLNIP-PDU, HEPP is absent.

- Information : the information field carries the CLNAP-PDU plus alignment header in the case of encapsulation and the CLNAP-SDU in the case of non-encapsulation.

2. For NNIs located within a single operator's domain, encapsulation of the CLNAP-PDU is an option. If the option is used, the encapsulation mechanism is identical with that just described.
   Alternatively, if the non-encapsulation option is used, the CLNIP-PDU remains identical with the CLNAP-PDU (transparent translation). However, if the non-encapsulating CLNIP-PDU subsequently crosses into a different operator's domain, the CLSF in the new domain will recognise (from the value of the PI field) that the received CLNIP-PDU is non-encapsulating. It will then map the non-encapsulating CLNIP-PDU into an encapsulating CLNIP-PDU before forwarding across the network towards the final CLSF on the path to the destination CPE. The mapping from non-encapsulating to encapsulating CLNIP-PDU is exactly as described for CLNAP-PDU encapsulation.

At the final CLSF (the second CLSF in the scenario shown), the received data is processed up through the protocol stack to recover the CLNIP-PDU. The CLL R&R function then operates on the CLNIP-PDU, performing a decapsulation function if necessary, to recover the original CLNAP-PDU. This is then routed to the destination CPE.

9.6.2.3 *Interworking between broadband connectionless data bearer service and switched multimegabit data service*

In this scenario, the BCDBS is provided within the B-ISDN by means of a connectionless service function, whilst the SMDS is provided by a MSS. Interworking between the two types of connectionless server involves the transparent transfer of connectionless service data units using the ATM switching capabilities of the B-ISDN. The network and protocol architectures for this scenario are illustrated in Fig. 9.25.

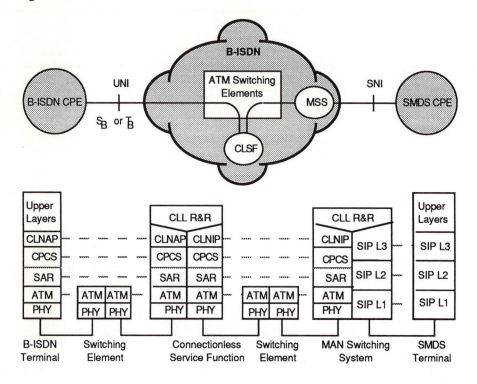

Fig. 9.25 *Interworking between BCDBS and SMDS*

As can be seen, a B-ISDN CPE gains access to the BCDBS by connecting to CLSF via ATM switching elements, whilst an SMDS-capable CPE is provided with direct access to a MAN Switching System (MSS). The B-ISDN CPE uses the CLNAP for communication with the CLSF; the SMDS CPE uses the SMDS Interface Protocol (SIP) suite for communication with the MSS.

In order for end-to-end connectionless communications to proceed between the two CPEs, it is clear that interworking must take place between the CLSF and the MSS. The interworking protocol used for this purpose is the CLNIP and encapsulation is a requirement between BCDBS and SMDS networks. At the CLSF, the CLNAP-PDUs received from the B-ISDN CPE are encapsulated into CLNIP-PDUs. The CLNIP-PDUs are then routed by means of the CLSF's CLL R&R function to the MSS, via the ATM switching elements as described previously.

At the MSS, the CLNIP-PDUs are received and decapsulated by the MSS's CLL R&R function to recover the CLNIP-SDU, which consists of the CLNAP-PDU plus alignment header. The alignment header is stripped off and the resulting data unit is passed to the SIP layer 3 entity. Here, the common PDU header and common PDU trailer are added to form the SIP L3-PDU. This is passed down through the protocol stack for transmission to the SMDS CPE. At the latter, the incoming data is transferred up through the protocol stack to recover the original upper layer protocols transmitted by the B-ISDN CPE. The process of CLNIP-PDU decapsulation and subsequent SIP L3-PDU encapsulation at the MSS is shown in Fig. 9.26.

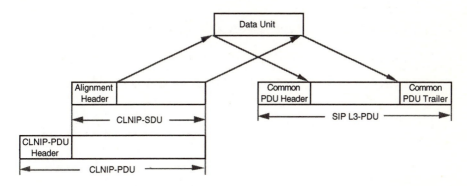

Fig. 9.26 *Decapsulation/encapsulation of BCDBS data units at MSS*

In the other direction of transmission, at the MSS the SIP L3-PDUs received from the SMDS CPE are decapsulated in the MSS's CLL R&R function. This process involves stripping off the common PDU trailer. The resulting data unit becomes the CLNIP-SDU, to which is added the CLNIP-PDU header to form the CLNIP-PDU. This is passed down through the protocol stack for transmission to the B-ISDN CPE. Here, the incoming data is transferred up through the protocol stack to recover the original upper layer protocols transmitted by the SMDS CPE. The process of SIP-L3-PDU decapsulation and subsequent CLNIP-PDU encapsulation at the MSS is shown in Fig. 9.27.

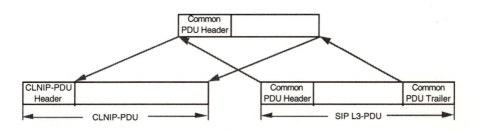

Fig. 9.27 *Decapsulation/encapsulation of SMDS data units at MSS.*

Although the fields comprising the data units to be exchanged at the interface between the two networks providing BCDBS and SMDS are common in terms of length and naming, their codings and interpretations differ on either side. For this reason, an Interworking Unit (IWU) is required at the network boundary to check and, if necessary, to modify the appropriate fields . This is to ensure that the fields in data units generated by the BCDBS and passed across the interface do not violate the coding rules for the fields in SMDS data units and vice versa. Such violations would result in data unit discard.

9.7 Bibliography

1. ITU-T Recommendation Q.922 : ISDN Data Link Specification for Frame Mode Bearer Services

2. ITU-T Recommendation I.233 : Frame Mode Bearer Services

3. ITU-T Recommendation I.555 : Frame Relaying Bearer Service Interworking

4. ITU-T Recommendation I.365.1 : Frame Relaying Service Specific Convergence Sublayer (FRSSCS)

5. ITU-T Recommendation I.363 : B-ISDN ATM Adaptation Layer (AAL) Specification

6. ETSI DE/NA 53205 DETS 'CBDS Over ATM : Framework and Protocol Specification at the UNI'

7. ETSI DE/NA 53206 DETS 'CBDS Over ATM : NNI Protocol Specification'

Chapter 10

Second generation LANs and MANs

Keith Caves

10.1 Introduction

In this chapter, we discuss a range of second generation Local Area Networks (LANs) and Metropolitan Area Networks (MANs). By 'second generation', we mean those LANs and MANs with protocols that permit operation at bit rates and with data throughput an order of magnitude higher than existing 'first generation' systems. In practice, this translates to a capability for operation at bit rates of at least 100 Mbit/s.

The following systems have been selected for discussion:

- Fibre Distributed Data Interface (FDDI)

- Distributed Queue Dual Bus (DQDB)

- Asynchronous Transfer Mode Ring (ATMR)

- Cyclic Reservation Multiple Access version II (CRMA-II).

These systems have been chosen as being reasonably representative of the second generation protocols currently available. FDDI is the first of the second generation LANs to be ratified as an international standard, while DQDB is well on the way to becoming the first of the MANs to be so ratified. The ATMR has been submitted to ISO for development as an international standard and for this reason alone would be worthy of investigation. CRMA-II has been included because, in the author's opinion, it possesses perhaps the greatest potential of the present crop of LAN/MAN protocols.

Although classed as a second generation LAN, FDDI is a development of the token ring first generation LAN protocol, albeit with important differences, and is consequently frame-based. The other systems are more 'modern' in the sense that they are capable of ATM cell-based operation.

The descriptions of the four systems will concentrate on their MAC protocols, which are vital to any insight and understanding of their operation. However, where appropriate, other attributes of the four will be explained for the sake of completeness.

10.2   Fibre Distributed Data Interface (FDDI)

10.2.1 Background

The term 'Fibre Distributed Data Interface - FDDI' was coined in the early 1980s by employees of the Sperry Corporation in the USA. Sperry had been instrumental in getting the American National Standards Institute (ANSI) subcommittee X3T9.5, responsible for data interfaces, to sponsor the development of FDDI. Logically, the IEEE would have been the proper choice of organisation to undertake such a standards development. However, at that time, the IEEE's remit only permitted development of LAN standards at data rates up to 20 Mbit/s. Then, the idea of a LAN operating over optical media at a data rate of 100 Mbit/s was revolutionary. Nowadays, the advent of LANs operating at speeds an order of magnitude greater than this is confidently awaited.

FDDI is a set of international standards. Basically, there are 4 individual standards which specify, respectively, the 4 major functional entities comprising FDDI. Together, these standards define a resilient data network operating at 100 Mbit/s over fibre optic media.

The MAC protocol for FDDI is based on the well known token ring protocol. Important differences, however, include the early token release mechanism for FDDI and its timed token rotation protocol, guaranteeing minimum bandwidth and maximum access delay. FDDI is based on a dual ring-of-trees topology and its legitimate subsets, including a dual ring without trees and a single tree. It enables flexible networks to be configured with path lengths ranging from a few metres up to 100 km, accommodating up to 500 stations. But note that even greater path lengths and numbers of stations are permitted so long as the default values of the timers used in the MAC protocol are recalculated.

10.2.2 Basic concepts

An FDDI system consists of a set of stations serially connected by optical transmission media to form a closed loop or ring. A station serves as the means of attaching one or more application devices, such as workstations or servers etc., to the ring for the purpose of communicating with other devices located elsewhere on the ring.

The physical topology of an FDDI network describes the interconnection of stations by means of duplex physical links; this topology forms a dual ring of trees. In a trunk ring, the arrangement of stations and physical connections forms a dual ring carrying contradirectional transmissions. In a tree, the duplex link provides one transmit and one receive path for one (normally the primary) of the dual rings. Note that various subsets of the dual ring of trees, including a dual ring without trees and a single tree, are also permitted.

As shown by the functional architecture given in Fig. 10.1 and indicated earlier, an FDDI station is defined by means of four functions.

(a) The Physical layer Medium Dependent (PMD) function resides in the lower sublayer of the physical layer of the OSI 7-layer reference model. PMD provides the driver/receiver requirements for the optical media, including electro-optic and opto-electric conversions. It also includes the optical media, connectors and optical bypass devices.

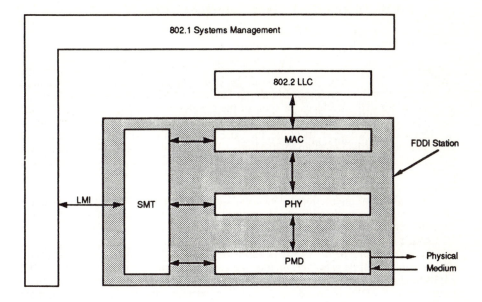

Fig. 10.1 *FDDI station functional architecture*

(b) The physical layer protocol (PHY) function resides in the upper sublayer of the physical layer. It provides the encoding/decoding, clocking and buffering requirements for transmission and reception of data onto and from the media, respectively.

(c) The Media Access Control (MAC) function resides in the lower sublayer of the data link layer. It provides stations with fair and deterministic access to the media. Its main concern is to offer service to the overlying LLC function to enable communication of frames of data, including frame transmission, frame repetition and frame reception.

(d) The Station Management (SMT) function manages the processes performed by the other three FDDI functions so that stations may work co-operatively on a ring. It communicates via an SMT management protocol with SMTs at other stations to provide distributed management of FDDI resources. It also communicates via the FDDI Layer Management Interface (LMI) to the system management process to enable the remote management of FDDI stations.

An FDDI station must possess a single SMT entity, but multiple PMDs, PHYs and MACs are possible. In addition, stations may be connected to the dual trunk FDDI ring or to trees connected to the trunk ring. Because of these two facts, a number of different station configurations result.

Dual attachment stations (DAS) form the FDDI trunk ring. A DAS must have two PMD/PHY pairs and at least one MAC entity, although an additional MAC allows the station to insert a MAC into both rings. A single attachment station (SAS) cannot connect directly to the trunk ring but only via a concentrator. The SAS needs a single PHY/PMD pair plus a single MAC.

Concentrators may be single or dual attachment devices or, more rarely, null attachment devices in the case of the root of a stand-alone tree. Concentrators are allowed to be MACless but if a MAC is provided (this is advisable) it must be connected in the logical ring downstream of any single attach devices being served.

Fig. 10.2 shows an FDDI dual ring consisting of a number of stations - dual attachment stations and dual attachment concentrators - interconnected by duplex trunk links. These effectively form two rings carrying contradirectional transmissions. The general arrangement is for the primary ring, comprising the set of stations connected by the primary links, to carry all the data transmissions, with the secondary ring on standby. Single attachment stations obtain their connection to the ring via dual attachment concentrators, but connect to only one ring (usually the primary). In this arrangement, the primary ring is made to go back and forth between the concentrator and its slave (single attachment) stations. Note that both single and dual attachment stations are interconnected by duplex fibre cable.

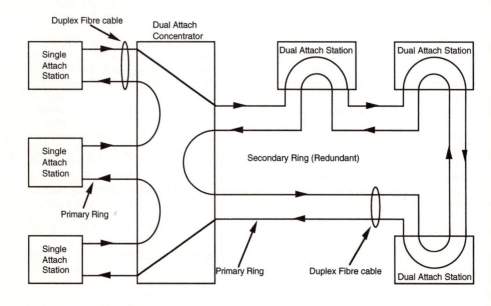

Fig. 10.2 *FDDI dual ring*

In the event of a catastrophic ring failure, e.g. caused by a break in the medium or a fault in a station, the stations on either side of the fault are able to perform a PHY 'wrapping' operation. This is a reconfiguration function that results in the primary (or secondary) ring input being looped back to the secondary (or primary) ring output to form a single operational ring from the still-active links of both primary and secondary rings. This property of FDDI resilience has one drawback in that the total fibre path length in the ring is now increased by approximately the length of the secondary ring.

10.2.3 Ring operation

Communication in FDDI takes place by means of the exchange of Protocol Data Units (PDUs), or frames of data, between MAC entities in the ring stations. These frames are constructed by a MAC on receipt of a request from either an LLC or SMT entity for the transmission of a Service Data Unit (SDU). Frames are generally queued by MAC to await transmission at the next opportunity for ring access. The MAC controls access of data frames onto the ring by use of the timed token rotation protocol. This requires a special PDU called the token - which is a unique sequence of data - to be passed around the ring from station to station.

A MAC wishing to transmit queued frames of data must first wait for the token to arrive. The token is then 'captured' by stripping it from the ring and replacing it with the frame(s) of data awaiting transmission. Immediately following completion of transmission of its data frame(s) the MAC issues a new token to give other stations the opportunity to access the media. The frames of data transmitted by the originating MAC are regenerated and repeated by the other active MACs on the ring. While repeating incoming frames, a MAC also examines the contained destination address for a match with its own address. If a match occurs, i.e. the frame is intended for the receiving station, the MAC copies the frame contents into its receive buffers as it transmits the frame forwards. Each frame of data transmitted eventually arrives back at the station that originated it, which is then responsible for stripping the frame from the ring. The originating station recognises the source address contained in the frame as its own address and strips the frame by replacing it with idle symbols.

MAC uses a timed token rotation protocol to control the transmission of frames of data onto the medium. Correct operation of the protocol depends on MAC maintaining two timers. The Token Rotation Timer (TRT) records the time of flight of the token in successive passes around the ring and is used to schedule frame transmissions. The Token Holding Timer (THT) is used for timing certain types of frame transmission, in conjunction with the Target Token Rotation Time (TTRT), a parameter determined during ring initialisation by the requirements of the most delay-sensitive application on the ring. In controlling and scheduling frame transmissions, the timed token rotation protocol provides two types of service, synchronous and asynchronous. Following ring initialisation and starting with the second token rotation, each MAC records in its TRT the time of flight of the token between successive passes around the ring. This is equivalent to maintaining a record of the amount of bandwidth used by ring stations during each token rotation. An early token, i.e. one which arrives back at a station before the value of TRT has reached TTRT, may be used for both synchronous and asynchronous transmissions. A late token, i.e. one which arrives after TRT has reached a value of TTRT, may be used for synchronous transmissions only.

The synchronous class of service, which is optional in FDDI, may be used for applications whose requirements for bandwidth and response time are known in advance. This permits the SMT at the stations serving these kinds of application to negotiate with an overall system management function for a pre-allocated amount of synchronous bandwidth. This is expressed as a proportion of TTRT; MAC is allowed to transmit synchronous frames for up to this length of time on each occasion that a usable token is captured. Of course, if there are no synchronous frames awaiting transmission, the resulting unused bandwidth is available for use by other stations.

The synchronous class of service, in addition to guaranteeing bandwidth, guarantees an average response time of not greater than TTRT and a maximum response time of not greater than twice TTRT. Note that response time means the elapsed time between a request for synchronous frame transmission and the arrival of a Token which permits its transmission onto the medium.

The asynchronous class of service, which is mandatory in FDDI, is meant for applications whose requirements for bandwidth are not predictable, i.e. are bursty, or whose response time requirements are not critical. Asynchronous bandwidth is allocated, when available, from the pool of bandwidth not pre-allocated to, or not currently being used by, synchronous applications. When a station has asynchronous data to transmit, it must capture an early token. The amount by which the token arrives early, i.e. (TTRT - TRT) is saved in the THT timer. TRT is then reset to time the next rotation of the token. THT reflects the amount of time available to this MAC for the transmission of asynchronous data. THT is enabled to count down during asynchronous transmissions and is disabled during synchronous transmissions. In the normal, non-restricted token mode of operation, the available asynchronous bandwidth is shared among all requesting MACs. In the optional restricted token mode, the asynchronous bandwidth is used for a single extended dialogue. A station wishing to enter restricted token mode first captures a non-restricted token.

It transmits its initial dialogue frame(s) and issues a restricted token. The addressed destination station(s) receive the initial frame(s) and enter restricted mode. Restricted tokens are then exchanged among all participating stations until the dialogue has been completed. The termination of restricted mode is made following transmission of the final dialogue frame(s), when a non-restricted token is issued. Note that stations with synchronous frames to send may also capture restricted tokens in order to complete their transmissions; only asynchronous transmissions are locked out.

10.2.4 FDDI Protocol Data Units

The token is a special PDU which is continually passed from station to station on the ring, conferring the right to transmit frames of data to a destination station. A description follows of the fields comprising the token as shown in Fig. 10.3.

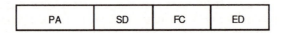

Fig. 10.3 *Token format*

• The preamble (PA) is always transmitted by the originating station as a minimum of 16 idle symbols. The length of the preamble may change as the token is repeated around the ring, due to PHY clocking requirements. However, any token with a PA greater than zero idle symbols must always be recognised and correctly processed.

• Starting Delimiter (SD) - a token is only considered as valid if it starts with the two symbols J,K.

• The Frame Control (FC) field defines the type of PDU, i.e. token.

• The Ending Delimiter (ED) for a token consists of a pair of T symbols.

A MAC PDU or frame is used for transmitting MAC, LLC and SMT messages to other ring stations. A description follows of the fields comprising the MAC frame as shown in Fig. 10.4.

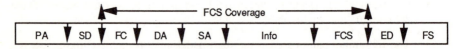

10.4 *Frame format*

• Preamble (PA) - the PA is always transmitted by the originating station as a minimum of 16 idle symbols. The clocking requirements at the PHYs of repeating stations may cause the PA length to be shortened or lengthened.

- Starting Delimiter (SD) - a frame is only considered valid if it starts with the two symbols J,K.

- Frame Control (FC) - defines the type of frame and associated control functions. It consists of two symbols conveying frame class (synchronous or asynchronous), frame address length (16 or 48 bit destination and source addresses) and frame format (the type of frame).

- Destination and Source Addresses (DA and SA) - each MAC frame carries a destination address and a source address in that order. MAC addresses may be either 48 bits or 16 bits in length, but all MACs must implement a 48 bit address capability. Further, a MAC operating with 48 bit addresses must be capable of functioning in a ring with other MACs operating with 16 bit addresses and vice versa.

- Information (INFO) - the INFO field carries zero or more data symbol pairs whose meaning is characterised by the FC field. The length of the INFO field is limited by the maximum PDU length of 9000 symbols.

- Frame Check Sequence (FCS) - used to detect the reception of erroneous frames, the FCS covers the FC, DA, SA, INFO and FCS fields. It is generated at the origin station by means of a standard 32-bit algorithm.

- Ending Delimiter (ED) - a single symbol T is used to indicate the end of a frame. However, ending delimiters and control indicators must be transmitted in pairs to maintain octet boundaries. This may require the addition of an extra T symbol following an even number of control indicator symbols.

- Frame Status (FS) - consists of three mandatory control indicators, used to indicate Error Detected (E), Address Recognised (A) and Frame Copied (C) conditions. Optional additional control indicators are permitted, defined by the implementer.

Peer MAC entities on an FFDI ring communicate by exchanging PDUs, i.e. frames of data . These frames are segmented by MAC into fixed length symbols which are passed across the MAC/PHY interface by means of service primitives. Each symbol received by PHY is encoded as a 5-bit code group for transmission onto the medium. Thus, 32 combinations of a 5-bit code group, i.e. 32 symbols, are available. The symbol set used by MAC consists of data and non-data symbols. A data symbol represents one quartet (4 bits) of binary data; 16 symbols are thus required to convey data. The 16 data symbols are denoted by the hexadecimal digits (0 -F); MAC only generates data symbols in pairs. The non-data symbols required are as follows:

- Q (Quiet), H (Halt) and I (Idle) symbols are used to convey line states. These are used to indicate the status of a physical link

- The JK symbol pair is used as the starting delimiter for a MAC frame. JK is a unique sequence, chosen because it can be recognised regardless of previously determined symbol boundaries, so that the start of frame may be established quickly and reliably.

- One or more T symbols constitute the ending delimiter for a MAC frame. However, the T symbol is not necessarily the final symbol in a frame transmission since it may be followed by one or more control indicator symbols. The rule is that a sequence consisting of ending delimiter followed by control indicators must comprise an even number of symbols. With an even number of control indicators, this means that a final T symbol must be added to the sequence.

- The R and S symbols are used as control indicators in MAC frames. They are used to indicate logical conditions associated with data frames, e.g. 'address recognised', 'frame copied'. The control indicators may be independently altered by repeating stations without changing the other fields of a frame. The S (Set) symbols indicates the logical 'on' or 'true' condition while the R (Reset) symbol carries the opposite meaning.

- The violation symbol V indicates detection of a non-valid symbol. Since there are 16 data symbols plus 8 non-data symbols, making a total of 24, this leaves a further 8 unused or invalid symbol assignments. The V symbol indicates a condition on the medium which is represented by one of these invalid code assignments.

10.2.5 FDDI PHY

As shown in Fig. 10.5, PHY is functionally organised as a receive part and a transmit part.

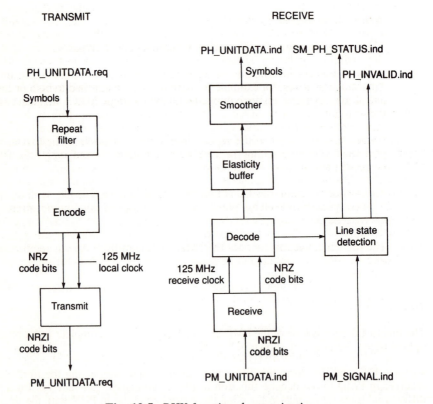

Fig. 10.5 *PHY functional organisation*

The receive part consists of 5 functions:

- The receive function derives a 125 MHz clock from the incoming NRZI bit stream received from PMD. It uses this clock in decoding the received NRZI to an equivalent NRZ bit stream. Both the NRZ bit stream and the recovered clock are presented to the decode function.

- The decode function decodes the incoming NRZ bit stream into an equivalent symbol stream. The decode function first detects a valid MAC frame start delimiter which it uses to establish the framing boundaries for the subsequent decoded symbol stream.

- The line state detection function determines the state of the inbound physical link. It notifies SMT of any changes in the received line state and reports receipt of any invalid symbols to the MAC.

- The elasticity buffer function is used to compensate for differences in clock rates at adjacent stations.

- The smoothing function compensates for the changes in length that occur to PDU preambles due to the actions of elasticity buffers.

The transmit part consists of 3 functions:

- The repeat filter function is necessary because, under certain circumstances, ring stations may be configured in such a way that there is no available MAC entity to process information being received on one of the incoming links. This situation arises, for example, in stations that are attached to both primary and secondary rings but are equipped with only a single MAC. In such cases, the symbol stream received on the MACless incoming link is repeated immediately, without processing, onto the appropriate outgoing link. By placing a repeat filter in the repeat path, invalid symbols or line states are prevented from propagating from the incoming link to the outgoing link, a task normally performed by a MAC entity.

- The encode function uses the autonomous local clock to encode symbols into 5 bit code groups which are presented as a 125 Mbaud NRZ output stream to the transmit function.

- The transmit function encodes the NRZ bit stream from the encode function into an NRZI bit stream and presents the latter to the PMD for transmission onto the optical medium.

10.2.6 FDDI PMD

The functional organisation of the PMD at a dual attach station, complete with optional optical bypass switches, is shown in Fig. 10.6.

In the example, the PHY/PMD interface uses DC coupled differential input/output signals. Rx+ and Rx- form the differential receive input connecting the fibre optic receiver in PMD to the decode function in PHY. Similarly, Tx+ and Tx- form the differential transmit outputs connecting the encode function in PHY with the optical transmitter in PMD.

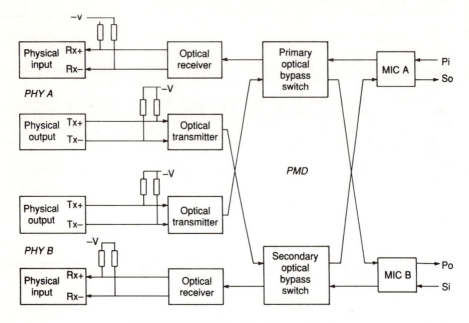

Fig. 10.6 *PMD functional organisation*

The functional entities comprising PMD are as follows:

- The optical transmitter accepts the differential NRZI digital baseband signals from PHY and produces at its output an equivalent stream of light pulses. An NRZI '1' is transmitted as light 'on' with NRZI '0' as light 'off'. The transmitter optical wavelength is 1300 nm nominal.

- The optical receiver accepts the optical signal received on the inbound physical link and converts this to the equivalent differential NRZI electrical signals at its output. As for the optical transmitter, the receiver operates at a nominal frequency of 1300 nm.

- The optical bypass switch is an optional feature for FDDI stations. With the station inserted into the ring, the optical signals from the inbound media propagate through the bypass switches to the optical receivers. Similarly, the outputs from the optical transmitters propagate through the bypass switches to appear on the outbound media. When the station is bypassed, the bypass switch causes each inbound medium to be connected to the appropriate outbound medium and each optical transmitter output to be looped back to the appropriate optical receiver input. In bypassed mode, the maximum attenuation through the bypass switch is 2.5 dB with a maximum switch operating time of 25 milliseconds.

- The Media Interface Connector (MIC) is the means by which an FDDI station attaches to the media.

10.3   Distributed Queue Dual Bus (DQDB)

10.3.1 Background

The IEEE committee 802.6 is responsible for the definition of the Distributed Queue Dual Bus (DQDB) standard. This is at an advanced stage of development and is expected to be ratified by ISO as an international standard in the near future.

DQDB defines a high speed, shared medium, distributed access protocol for use in subnetworks of a Metropolitan Area Network (MAN). It is intended for the support of integrated multiservices communications. However, the current version of the standard specifies the provision of a connectionless (CNL) data service to the LLC sublayer. Whilst additional functions describing support for an isochronous service and a connection-oriented data service are included, their presence is for tutorial purposes and they are not a mandatory part of the standard.

The Physical Layer specified permits the use of different transmission systems by means of appropriate Physical Layer Convergence Protocols (PLCPs). The current version of the standard includes PLCPs for :

- ANSI DS3 (47.736 Mbit/s)

- CCITT G.703 (2.048 Mbit/s, 34.368 Mbit/s and 139.264 Mbit/s)

- CCITT synchronous digital hierarchy (155.520 Mbit/s).

10.3.2 Basic concepts

A DQDB subnetwork consists of two unidirectional buses connecting a number of nodes, as shown in Fig. 10.7. The two buses, conventionally called bus A and bus B, carry contradirectional transmissions; by this means, full duplex communications between any pair of nodes on the subnetwork is enabled.

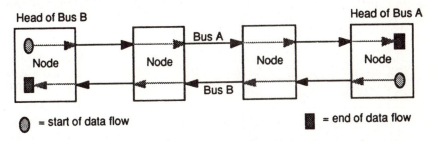

Fig. 10.7 *Open dual bus architecture*

For efficient communications, it is desirable for a node to know on which bus to transmit to reach a desired destination node. Basically, this can be arranged in one of two ways:

- by learning: e.g. by sending data on both buses and observing on which bus a response is received

- by pre-loading node addresses plus their 'transmit' bus from a system management function.

As shown in Fig. 10.8, a looped bus architecture is also possible which differs from the open bus only in that the end points of the bus are co-located. Note, however, that although the topology is physically a ring, logically it is still a bus. This means that the data does not flow from the end of the bus through to the beginning of the bus but is terminated exactly as for the open bus.

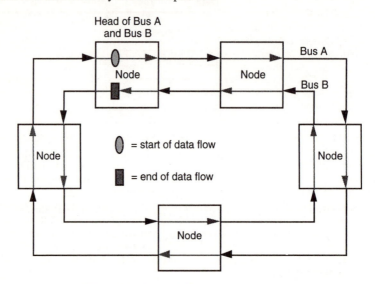

Fig. 10.8 *Looped bus architecture*

The looped bus architecture conveys an advantage in that it permits network reconfiguration in the event of a bus fault. For example, a physical break in the medium would cause the two nodes on either side of the fault to assume head of bus A and head of bus B responsibilities, respectively. In effect, the head of bus functions are moved out of the single node in the undamaged system into two nodes in the faulty system. Note that the looped bus has thus reverted to the open dual bus architecture.

The node at the head of each bus is responsible for generating data onto the bus . This data can be in the form of DQDB layer management information octets or 53 octet fixed length slots. The management information ensures the correct operation of the subnetwork. The slots are used to carry data between communicating nodes.

An access unit in each node enables the node to read from each bus and write to each bus, under control of the DQDB access protocol. Data reaching the end of the bus is terminated there.

The DQDB layer (equivalent to the MAC) provides two modes of access to each bus. These are:

- Queued Arbitrated (QA) access, which uses QA slots under control of the distributed queue protocol to provide, typically, connectionless data service

- Pre-Arbitrated (PA) access, which uses PA slots under control of the PA protocol to provide, typically, isochronous services.

A node on the DQDB subnetwork consists of an Access Unit (AU) plus the means for attaching the AU to each of the buses, as shown in Fig. 10.9. The AU is responsible for performing the MAC layer functions, including the DQDB access protocol.

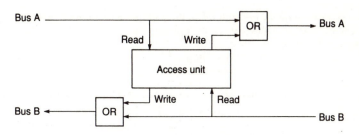

Fig. 10.9 *Access unit attachment*

The AU attaches to each bus by means of one read and one write connection. In order to write data onto the bus, a logical OR function is performed on the data from upstream with the node's own data. Data is read from the bus logically ahead of the OR function used for writing to the bus. This is so that the read data remains unaffected by any write operations being performed by the node.

The read operation allows nodes on the bus to copy data passing by but not to remove it. Data on the bus can only be altered when permitted by the access protocol.

The read and write functions may be implemented by either passive or active means. However, it should be noted that, when using optical fibre for the bus media, passive techniques limit the number of nodes on the bus because of optical power budget limitations.

10.3.3 DQDB slots

The 53-octet slot, shown in Fig. 10.10, is the basic unit of data transfer between nodes on a subnetwork. Every slot contains a 1-octet Access Control Field (ACF) plus a 52-octet segment.

Access control field (1 octet)	Segment (52 octets)		

- Busy (1 bit)
- SL-type (1 bit)
- PSR (1 bit)
- Reserved (2 bits)
- Request (3 bits)

Busy	SL-type	Slot state
0	0	Empty, QA slot
0	1	Reserved
1	0	Busy, QA slot
1	1	PA slot

Fig. 10.10 *DQDB slot access control field*

The ACF contains a number of sub-fields consisting of 1, 2 or 3 bits. In particular, the Busy, SL-Type and Request sub-fields are manipulated by the DQDB access protocol to control access by a node's data to the slots that pass by on the two buses.

The Busy bit indicates whether the slot is empty (=0) or occupied by a node's data (=1). The SL-Type bit indicates either a QA (=0) or a PA (=1) slot. The combinations of Busy and SL-Type bits have particular meanings as shown in Fig. 10.10.

The Request field contains the three REQ bits. These are used in the distributed queue access protocol, one bit for each of three priority levels. A REQ bit is set by a node transmitting on bus A or B to request access to a QA slot on bus B or A, respectively. The particular REQ bit set (REQ-2, REQ-1 or REQ-0) indicates the priority level of the associated request (from high to low, respectively).

All the bits of a QA slot are set to 0 by a slot marking function at the Head of the Bus (HOB). The Busy and SL-Type bits set to 0 in the ACF indicate to nodes downstream from the HOB that the slot is an empty QA slot. When a node's AU gains access to a QA slot to transfer information, it marks the QA slot by setting the Busy bit to 1 and writes into the segment field.

PA slots are generated by the HOB slot marking function with the Busy and SL-Type bits of the ACF set to 1, with all other ACF bits set to 0. The Busy and SL-Type bits of a PA slot remain unchanged on the bus whereas the REQ bits may be changed in accordance with the rules for the distributed queue.

A slot segment carries the data that is transferred in the slot. Each segment contains a 4 octet header plus a 48 octet payload, as shown in Fig. 10.11.

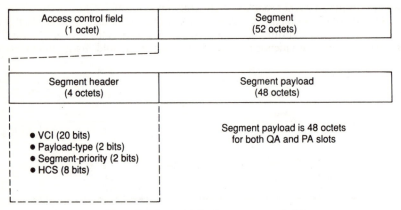

Fig. 10.11 *DQDB slot segment field*

The segment header consists of 4 fields. The 20 bit VCI field identifies the virtual channel with which the segment is associated. The 2-bit payload-type field is used to denote the type of information carried - 00 indicates user data. The segment priority field of 2 bits is reserved for future use with multiport bridging. The 8-bit header check sequence provides a single bit correction capability and a multi-bit detection capability for errors occurring in the segment header.

A QA segment is generated as all zeros by the HOB node. All the fields of the header are inserted by the AU that also writes data into the QA segment payload of an empty slot.

The HOB slot marking function writes the header for a PA segment. It also sets all the bits of the segment payload to zero. The PA segment header remains unchanged as the slot passes along the bus. The PA segment payload consists of 48 isochronous octets which may be shared among a number of isochronous service users. By generating PA segments at an 8 kHz rate, the HOB function can ensure an isochronous service capability for PCM voice samples.

10.3.4 Distributed queue access protocol

The distributed queue access protocol is the means by which fair and deterministic access to QA slots on the two buses is provided for nodes that request service. In particular, the protocol is aimed at connectionless (CNL) data services that are bursty in nature.

The protocol uses the BUSY and REQ bits in the ACF of QA slots. When a node queues a segment for access to a bus, it informs other nodes of this by setting a REQ bit in the next available QA slot on the backward bus. By counting such requests, but subtracting the number of free slots (BUSY bit =0) which pass by and fill the requests, a node keeps track of the number of segments queued awaiting transmission. Further, this enables the node to determine the number of outstanding requests queued ahead of its own request and, hence, how many free slots to let pass before using one for its own segment. This operation is equivalent to maintaining a single distributed queue for the network to which all nodes have access.

A priority mechanism is operated by means of separate queues for the three priority levels. Priority is always given to segments queued in higher level queues, but it should be noted that any CNL data segments must be sent at the lowest priority. The use of priorities for other services is presently under study.

Fig. 10.12 shows a part of a subnetwork with nodes attached to bus A and bus B. In considering the algorithm for access to bus A, bus A is the forward bus and bus B is the reverse bus. An identical but inverse arrangement holds for access to bus B.

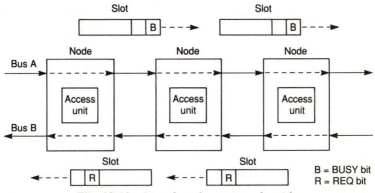

Fig. 10.12 *Distributed queuing algorithm*

As indicated previously, the ACF in each slot on the two buses contains a BUSY bit and a REQUEST field of 3 REQ bits. The BUSY bit indicates whether the slot is in use or free and the REQ bits indicate when a QA segment has been queued to send.

When an AU has a segment to transmit on the forward bus, it writes a REQ of the appropriate priority level in the next free REQ bit passing on the reverse bus. The REQ bit informs all nodes upstream on the forward bus that a new segment has been queued to send at a particular priority level. For each AU, the access protocol allows only one segment to be queued at each priority level on the two buses - a maximum of 6 segments.

Fig. 10.13 shows the actions taking place at a node which currently has no segments queued to send.

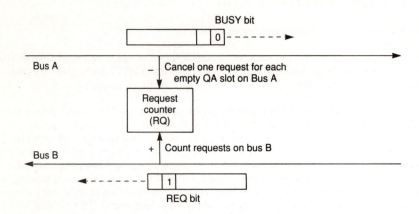

Fig. 10.13 *QA access protocol at node without segments queued*

The AU in the node keeps track of the number of QA segments queued downstream on the forward bus by counting REQ bits that pass on the reverse bus. This count is held in the AU's request counter (RQ), which is incremented each time a REQ bit passes on the reverse bus. However, the RQ count is decremented for each free slot (busy bit =0) that passes by on the forward bus, since this slot will be filled by one of the segments queued for access in a downstream node.

By means of these two actions, of incrementing and decrementing a counter with requests on the reverse bus and free slots on the forward bus, respectively, a dynamic record is maintained of segments queued to send downstream of this node.

Fig. 10.14 illustrates the actions at a node with a segment queued to send on the forward bus.

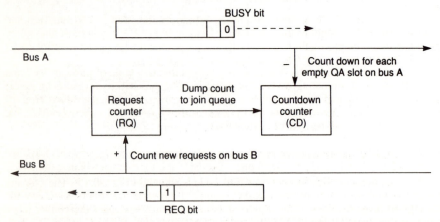

Fig. 10.14 *QA access protocol at node with segments queued*

Having issued a request (by writing a REQ bit) on the reverse bus the AU transfers the current count held in the request counter to a countdown counter (CD). The RQ counter is then reset to zero and CD now holds a record of the number of segments queued ahead of the node's own segment for access to the forward bus. This action effectively places the node's segment in the distributed queue.

Now, every time a free slot passes by on the forward bus the CD counter is decremented by one. As soon as the CD count reaches zero, the AU may transmit its queued segment in the next available free slot.

Whilst the AU is awaiting its turn to transmit a segment, REQ bits passing by on the reverse bus continue to be counted in the RQ counter. In other words. the number of segments newly queued for access is still tracked by the RQ counter during the operation of the CD counter.

By means of these two counters in each AU, one counting as-yet unsatisfied access requests and the other counting down before access, a distributed queue is established and maintained to control access to the forward bus.

The basic QA access protocol exhibits a degree of unfairness of access to the bus which is dependent on a node's position relative to the HOB. Those nodes nearer to the HOB may receive a proportionately larger share of available bandwidth than those nodes further away, under periods of high demand for service. To counteract this tendency, a bandwidth balancing mechanism has been introduced. This requires nodes occasionally to skip the use of an empty QA slot. The requirement is implemented by means of a bandwidth balancing (BWB) counter for each bus in every node.

Whenever a node transmits a QA segment, the BWB counter associated with that bus is incremented. At some predetermined count, the BWB counter rolls over to zero. When this occurs, it provides an indication that the AU should ignore the next free QA slot to pass on that bus. Effectively, this is equivalent to incrementing the RQ counter for that bus when no QA segment is queued, or incrementing the CD counter for that bus when a QA segment is queued.

PA slots, used typically for isochronous data transfer, are accessed differently to the QA slots. PA slots are designated by the node at the head of bus and generated in a periodic manner, e.g. at a regular 8kHz rate for transfer at 64 kbit/s digital PCM voice information.

The payload of a PA segment consists of 48 octets; these are available for use by different AUs on the bus. Thus, an AU may write zero, one or more octets of a PA segment.

The VCI field in the segment header is written by the head of bus function and identifies the PA segment to the AUs. AUs wishing to use PA octets must also know the offset of the octet(s) with respect to the start of segment. This octet identity - VCI plus offset - is conveyed by layer management procedures.

To read and write into PA octets, an AU will maintain a table of relevant VCI and octet offsets for those octets 'owned' by that AU. The AU will then read from octets marked for reading and write to octets marked for writing. Note that reading from (or writing to) a given octet occurs for the duration of a call, following which the AU table entry for that octet will be erased.

10.3.5 DQDB node architecture and services

The functional architecture of a DQDB node is shown in Fig. 10.15. The QA and PA access control functions are used by a range of convergence functions to provide the DQDB layer services. The DQDB standard presently specifies the convergence function that provides the ISO MAC sublayer service to the ISO LLC sublayer. Also specified in a separate part of the standard is an isochronous service, whilst a connection-oriented data service may be defined in the future.

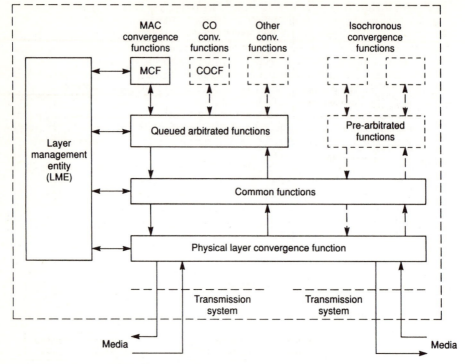

Fig. 10.15 *DQDB node functional architecture*

The convergence entities shown in the dotted boxes are not included in the current standard description but are shown for completeness.

Convergence functions residing at the same node are not permitted simultaneous access to a slot with the same VCI. The VCI may thus be used to direct a slot segment to the appropriate convergence function.

The DQDB standard defines a number of different physical layers. Each definition comprises two protocol functions:

- a transmission system function that describes the characteristics of the inter - nodal links

- a physical layer convergence function that adapts the capabilities of the transmission system to the requirements of the DQDB layer.

The physical layer convergence function uses a Physical Layer Convergence Procedure (PLCP). The PLCP performs a mapping from the slot octets and management information octets into a format that is suitable for transfer by the underlying transmission system.

A unique PLCP is required for use with each different transmission system. In the event that the transmission system already provides the desired physical layer service characteristics, the PLCP will be null.

The provision of MAC service to LLC consists of the origin node segmenting the MAC service data unit (MSDU) received from LLC into fixed length units and transferring these units to the destination node, which reassembles them into the original MSDU. The segmentation process is illustrated in Fig. 10.16.

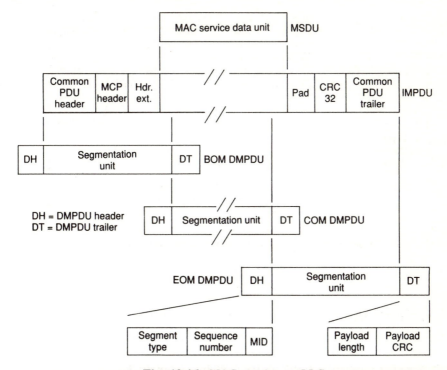

Fig. 10.16 *MAC service to LLC*

Prior to the segmentation process, an initial MAC Protocol Data Unit (IMPDU) is formed. This consists of the original MSDU plus an IMPDU header (comprising common PDU header and MAC Convergence Protocol (MCP) header), an optional header extension, an optional 32-bit CRC, a common PDU trailer and a variable length PAD field to ensure 32-bit alignment of the fields added to the MSDU.

The IMPDU is now segmented into fixed length (44 octets) segmentation units. To each of these is added a header (containing segment type, sequence number and Message Identifier (MID) fields) plus a trailer (containing payload length and payload CRC fields) to form the Derived MAC Protocol Data Units (DMPDUs). The DMPDU is the unit of data that is transferred by means of the 48-octet QA segment payload. Note that DMPDUs are labelled by means of the segment type field in the header as either Beginning of Message (BOM), Continuation of Message (COM) or End of Message (EOM).

10.4 Asynchronous Transfer Mode Ring

10.4.1 Background

The Asynchronous Transfer Mode Ring (ATMR) proposed standard has been submitted to ISO by the Japanese national body for development as an international standard.

The protocol is based on BT's Orwell system, but has been enhanced and developed to improve its performance. It is a high speed, shared medium access protocol designed to operate on a dual slotted ring. It uses the standard CCITT B-ISDN ATM cells, containing 53 octets with 5 octets of header and 48 octets of payload, as the ring slots. The slots provide the transport mechanism for the support of multimedia services.

The physical medium defined in the proposal is the standard SDH STM-4 transmission system designed for operation at 622.08 Mbit/s as specified in CCITT Recommendations G.707, G.708 and G.709. The ATMR cells are mapped into virtual containers (VC-4) carried in STM-4 in the standard manner as specified in CCITT Recommendation I.432.

10.4.2 Basic concepts

The ATMR protocol provides fully distributed access control to the dual slotted ring such that nodes may transmit data concurrently. This is made possible by the short (53 octet) slot lengths.

In operation, the system requires a master node which is determined at system initialisation. The master performs a number of functions, including:

- provision of a timing reference by which means the ATMR network achieves synchronisation. The master will use a timing source from an external network, provided that a connection to a suitable network exists; otherwise, the master's node clock is used.

- indication to the other nodes on the network of which ring is the active ring.

- marking of cells passing by on the active ring by setting the Monitor bit (M-bit). Cells arriving at their destinations then have the M-bit reset, so that any cell that arrives back at the master with its M-bit set is treated as a garbage cell and released, i.e. the address field set to all zeros. Note that a separate monitor node could be identified to perform this cell monitor function, but in practice the master node is the obvious choice.

The ATMR cell is the basic PDU used for the transport of data between ring nodes. This cell is identical with the standard CCITT B-ISDN ATM cell. The ATMR protocol offers asynchronous access to the cell stream passing by on the active ring. It is thus well suited to the support of non-isochronous data. It can also be used to support delay-sensitive traffic such as voice and video, by means of a suitable convergence function to manage the buffering and retiming processes inherent in isochronous service provision.

The ATMR protocol can guarantee bandwidth to nodes by virtue of its window size and reset mechanisms. These also ensure that access delays are tightly bounded - the mechanisms are explained later.

The ATMR system consists of a number of Access Nodes (ANs) interconnected by duplex media to form a dual closed loop or ring. In operation, one of the rings is designated as active and the other as standby; the two rings carry contradirectional transmissions. Fig. 10.17 illustrates the system configuration for an ATM ring.

The ANs provide the means of access to the ring for terminals, other LANs, etc. A single Access Unit (AU) within each node operates the ATMR protocol and provides the functionality for attaching to the underlying transmission system.

In common with other dual ring systems, the ATMR provides a loopback capability such that node or medium failure can be temporarily cured. The drawback with loopback operation is that the total medium path length is effectively doubled, although this is not so potentially serious with the current protocol as with, say, FDDI.

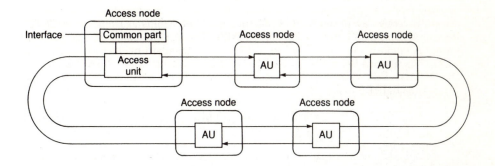

Fig. 10.17 *ATMR system configuration*

10.4.3 ATMR node architecture

The ATMR node functional architecture is shown in Fig. 10.18. It consists of a number of functions which fall into 3 layers.

The MAC convergence entity, which handles 'normal' LAN connectionless data, is part of the ATM adaptation layer. Other convergence entities shown in the lightly outlined boxes are not part of this proposed ATMR standard but are shown for completeness.

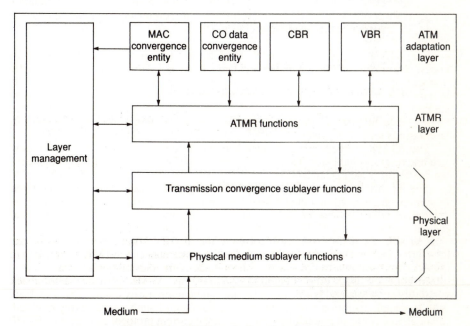

Fig 10.18 *ATMR node functional architecture*

The MAC convergence entity performs the following functions:

- it receives a PDU from the overlying LLC sublayer as a MAC SDU (MSDU).

- it creates an Initial MAC Protocol Data Unit (IMPDU) from the MSDU by adding control information plus an IMPDU header and trailer.

- it segments the IMPDU into fixed length units from which it creates Derived MAC PDUs (DMPDUs) by adding control information plus a DMPDU header and trailer.

- DMPDUs are passed to the ATMR layer as ATMR_SDUs in ATMR_DATA.request primitives.

A mirror image set of functions is performed when accepting ATMR_DATA.indication primitives from the ATMR layer and presenting MSDUs to the LLC sublayer. Note that the above MSDU segmentation and reassembly mechanisms are as described for the ISO/IEC DIS 8802-6 definition of the DQDB protocol.

The ATMR functions are part of the ATMR layer, which is equivalent to the ATM layer of the B-ISDN standards defined by the ITU. These functions include:

- the distributed window size and reset mechanisms to ensure fairness of access to the ring and bounded access delays.

- handling of multiple priority classes for different QOS requirements.

- elimination of garbage cells by use of the MT bit.

The transmission convergence sublayer entity is part of the physical layer. Its functions include:

- provision of Master node functions.

- performance of Master node contention process.

- configuration control to ensure that all ATMR node resources are correctly configured into an operational ring, both at network initialisation and following ring failure.

- transmission convergence functions, to ensure correct mapping of ATMR cells onto the underlying transmission system.

10.4.4 ATMR cells

The ATMR cell is the basic protocol data unit at the ATMR layer. All cells are transferred between adjacent nodes on a unidirectional ring. As shown in Fig. 10.19, the ATMR cell consists of a 5-octet header and a 48-octet payload. The order of transmission on the physical medium starts from the 1st octet with the MSB of that octet - bit 8 - transmitted first.

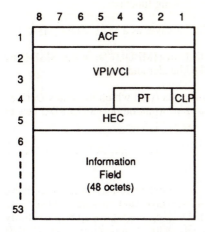

Fig. 10.19 *ATMR cell format*

The Access Control Field (ACF) consists of 8 bits. Its purpose is threefold;

(i) It carries a reset signal for window counters and timers in node Access Units (AUs). Three types of reset signals are available as will be explained later.

(ii) It indicates the so-called 'busy address' of an AU. When an AU has cells to transmit and is not prevented by the ATMR protocol from transmitting them, it overwrites its Busy Address (BA) in every incoming cell except reset cells.

(iii) It indicates the monitor state of the cell by use of the Monitor bit (M-bit). This is the MSB of the ACF.

Note that the BA field can indicate a maximum of 125 addresses since the MSB is used for the M-bit and the reset signals require 3 of the remaining ACF combinations. This is clearly a limitation, since each AU normally has its own busy address. To circumvent this limitation, one of the address codes (X0000000) is reserved for use as a common busy address. In appropriate situations, this may be allocated to multiple AUs so that AUs with unique BAs co-exist with AUs with common BAs. However, the latter suffer a restriction in being unable to issue reset cells.

The Virtual Path Identifier/Virtual Channel Identifier (VPI/VCI) field is 20 bits long and carries the ATMR cell routing indication. Routing can be based on either a flat or a hierarchical addressing structure.

The Payload Type (PT) field is 3 bits long and is identical to that of the standard CCITT ATM cell. It indicates the nature of the information field.

The Cell Loss Priority (CLP) bit indicates whether the cell may be discarded by the network under congestion conditions, i.e. it is a QOS indicator. It is used at a bridge between the ATMR network and another network for congestion control.

The Header Error Control (HEC) field is 8 bits long. It is used for checking the ATMR cell header for errors and also for cell delineation. The CRC polynomial is identical with that specified in CCITT B-ISDN standards.

10.4.5 ATMR access protocol

The ATMR protocol provides fully distributed access control for slotted ring networks. The ring architecture ensures location-independent access for nodes. The short, fixed (53 octet) slot size enables multiple AUs to access the ring concurrently. The ATMR protocol is relatively insensitive to ring size owing to its distributed access control and concurrent access features. The protocol is also bit-rate independent and scaleable. All service types - voice, video, data - are supported by the slot transfer mechanism. Since the slot is identical with the CCITT-defined ATM cell, compatibility with B-ISDN is assured. Statistical multiplexing of the different traffic types leads to efficient use of bandwidth with low delay and high throughput. Two priority levels are defined for classes of traffic with different Quality of Service (QOS), i.e. throughput and delay requirements.

In order to ensure fairness of access for nodes to the available ring bandwidth, each node is given a window size which limits the number of slots it may transmit in a given time period. A distributed monitoring system is used to determine when to restart transmission from all nodes that have transmitted their current window size. Transmitted slots are released (freed) on reaching their destinations. This leads to spatial reuse, so that slots can be used more than once within one rotation of the ring, which is equivalent to a bandwidth multiplication mechanism.

At the time of ring initialisation, a management protocol is used by all nodes to perform the master node contention function. The basis for this is a node's master node Ppriority value and its physical address. The node with the highest priority and largest numerical address becomes the master node. The master then generates a stream of empty slots which are regenerated at each node in turn, eventually arriving back at the master. This effectively initialises the ring, following which ring traffic may be supported.

A node with information queued to send, and which is not currently prohibited by the ATMR protocol from sending, transmits cells onto the ring to occupy empty passing slots. All slots passing by on the ring are examined by each node, where the value of the VPI/VCI field is checked. If the value of this address matches the node's own address, the contents of the slot are copied into the node's receive buffers. The value of the VPI/VCI field is then set to zero to effectively release the slot. The empty slot is then either relayed downstream for use by some other node or, if this node has information queued in its transmit buffers, is used for its own transmission.

If the value of the VPI/VCI field of an incoming slot does not match the node's own address, the slot is relayed downstream with the VPI/VCI unchanged.

Fairness of access to the medium is achieved by means of the window size and reset mechanism implemented in each AU. An AU's Window Size (WS) is a pre-assigned number determined by an overall system management function. The WS represents the number of cells that an AU is allowed to transmit before either issuing or receiving a reset cell. The ATM ring access control mechanism is described in the following, in conjunction with Fig. 10.20.

Each AU has a Window Counter (WC) whose initial value is set to the node's allocated WS. Thenceforth, WC is decremented by one every time that the AU transmits a cell until the value of WC reaches zero. At this time, the AU must stop sourcing further cells. Eventually, all the AUs on the ring will stop transmitting cells, either because they have sent their quota (WC=0) or because they have no information queued in their transmit buffers. The AUs then depend on receiving a reset cell in order to reset their WCs and restart transmissions. Note that the value of WS may be different for each AU, since it will be set at some value appropriate to the applications served by the particular node.

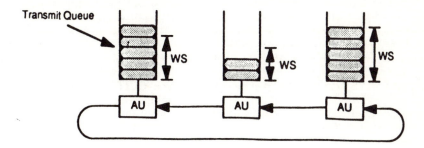

Fig. 10.20 *ATMR access control mechanism*

An active AU on the ATM ring is defined as one which has the right to transmit cells (WC>0) and which also has at least one cell queued to send. Such an active AU always overwrites its own Busy Address (BA) in the ACF of the cell header of a passing cell to indicate that it is still active. An inactive AU, on the other hand, is defined as one which does not have the right to transmit cells (WC=0) or which has no cells queued to send. Such an inactive AU does not overwrite its BA in the ACF of passing slots.

When an active AU recognises its own BA in the ACF of an arriving slot, it concludes that all other AUs on the ring have become inactive, since its BA has not been overwritten. Then, as soon as this last active AU becomes inactive on completing its quota (or on running out of cells to send), it issues a reset cell. It also resets its window counter to the value of its WS. The reset cell circulates around the ring causing each AU in turn to reset its window counter to its individual WS value. This enables AUs with cells ready to transmit to revert to the active state and to begin sending cells again.

The interval between reset cells is the reset period. Since there is a finite number of AUs, each with a finite value of WS, there is an upper bound on the reset period. This means access delays can be bounded and bandwidth guaranteed to AUs. Moreover, if AUs do not have cells to transmit during a particular reset period, that period becomes shorter, effectively enabling other AUs to share the unused bandwidth.

In the ATMR protocol, the window size and reset mechanism can be extended to cover multiple Priority levels. A separate window size is allocated to each priority class in accordance with the QOS requirements (throughput, delay) of the associated applications. For example, in a system operating two priority levels, within a given reset period any higher priority Class_1 cells are always transmitted ahead of lower priority Class_2 cells. Higher priority cell transmission continues until either WC=0 or until no Class_1 cells remain; Class_2 transmission can then begin.

10.5   Cyclic Reservation Multiple Access   (Version II)

10.5.1 Background

The Cyclic Reservation Multiple Access (CRMA-II) MAC protocol has been proposed by IBM researchers for use in high performance LANs and MANs. Designed to operate at Gbits/s data rates, it is capable of supporting multimedia services on both bus and ring topologies.

The protocol uses a slotted transmission structure to provide concurrent access for multiple users. The slots are designed to accommodate a single ATM cell. Slot delineation is performed by start and end delimiters. The protocol caters for any combination of frame-based LAN traffic and cell-based ATM traffic. Frames of data are carried in contiguous slots, with a node's insertion buffer holding slots arriving from upstream whilst transmission is in progress. Transmission of data frames in contiguous slots greatly reduces delays and removes the need for segmentation and reassembly functions at nodes.

10.5.2 Basic concepts

The slots used to carry information are embedded between start and end delimiter pairs. Whole frames of data are segmented into slot-length payloads but are then transported in contiguous slots. However, these contiguous frame-carrying slots have a start delimiter, but only the final slot terminates with an end delimiter. For ATM cells, each of which is transmitted within a single slot, both start and end delimiters are required.

Just as slots nominally carry both a start and an end delimiter, so do they carry source and destination address fields. Again, in the case of frames of data transported in contiguous slots, only the first slot carries the address fields. These measures minimise overheads and contribute to system throughput.

A single bit field in the start delimiter indicates whether a slot can be freed (emptied of data) at the source or at the destination. Source deletion is the norm for multicast slots whereas destination deletion generally applies to data addressed to a single node. Destination deletion and subsequent reuse of the slot by either the deleting node or one further downstream effectively increases data throughput to well in excess of the medium bit-rate.

A single bit busy/free field in the start delimiter is used by the access protocol. This provides fast access to slots under low to medium traffic loads. Under heavier loads, a reservation mechanism ensures sustained high network utilisation, fair access to slots and bounded access delays.

On a ring, a scheduler node is appointed to be responsible for both the reservation mechanism and the monitor function. The latter ensures that busy slots not freed by the deleting nodes (e.g. due to address corruption) are freed in the monitor function. This takes place by means of a monitor bit in the start delimiter, which is set by the monitor function and reset by a deleting node. Thus, no busy slot should ever reach the monitor with its monitor bit set unless data corruption or other fault has occurred.

10.5.3 CRMA-II slots

The CRMA-II slot is the basic unit of information transfer between nodes on a subnetwork. Slots are nominally of constant length, although, as explained previously, concatenated slots may lack end delimiters (except the last slot) and address fields (except the first slot).

The format of a slot is shown in Fig. 10.21 and consists of a number of 32-bit words. The start delimiter word comprises two fields, a single-byte synchronisation field and a 3-byte slot control field. The latter carries a number of parameters:

– slot type, denoting asynchronous, synchronous or isochronous (only asynchronous slots considered here)

– priority, defining the quality of service associated with the slot

- busy/free, denoting the state of the slot

- gratis/reserved, denoting (for free slots) which slots are available for unlimited access and which require prior reservation

- first/middle/last/only, denoting the position of the slot in a multi-slot frame or, in the case of 'only', denoting a single slot transmission (e.g. ATM cell)

- source/destination deletion, indicating which node is responsible for freeing the slot

- monitor, indicating whether the slot has been marked by the monitor function.

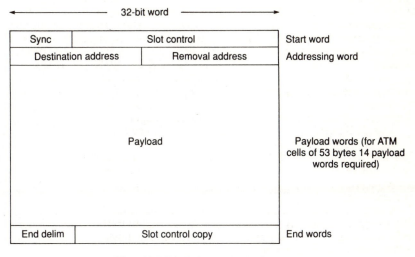

Fig. 10.21 *CRMA-II slot format*

The addressing word contains a 16-bit destination address plus a 16-bit source address.

The payload consists of a number of consecutive payload words. This number can be chosen at initialisation time to best suit the applications being supported; to carry ATM cells, however, 14 payload words are required.

Finally, the slot is terminated with an end delimiter word. This carries a single byte end delimiter field plus a copy of the slot control field carried by the start delimiter word. This redundancy provides an error detection mechanism.

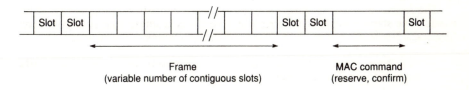

Fig. 10.22 *Typical slot sequence*

Fig. 10.22 shows a typical slot sequence. This contains individual slots, a multi-slot, or variable number of contiguous slots carrying a complete frame of data, and a MAC command, also variable in size as will be explained later.

10.5.4 CRMA-II access protocol

The CRMA-II access protocol provides fair and deterministic access to slots passing by on the medium for nodes with information queued to send. In the following, the protocol will be described in the context of a connectionless data service, although the same principles apply to other service categories.

In CRMA-II, the Start delimiter of a slot carries busy/free and gratis/reserved fields. These enable slot access through two very distinct mechanisms:

- immediate access to gratis slots

- access to slots reserved during a prior reservation mechanism.

The gratis/reserved field distinguishes slots in these two states. In both cases, the busy/free field must denote that the slot is free for access to be permitted.

Since it is based only on a busy/free decision, fast access is obtained to gratis slots. Reserved slots, however, are only created by the scheduler node following the prior exchange of reserve and confirm MAC commands between the scheduler and all other nodes on the ring. These MAC commands are not part of a slot but are self-contained entities carried between start/end delimiter pairs. Busy slots - whether gratis or reserved - become free gratis slots when freed by the deleting node and so available for re-use.

Under heavy traffic loads, gratis access of slots occurs mainly through spatial re-use following destination deletion. Under these conditions, the reservation mechanism ensures that more reserved slots are allocated to nodes with fewer gratis-access opportunities. Fairness and tightly bounded access delays are thus maintained.

The configuration and operation of a CRMA-II system throughout the following will be described in terms of a single ring topology, as illustrated in Fig. 10.23. However, the protocol is straightforwardly extensible to dual rings and dual bus topologies also. The system thus consists of a number of nodes connected by

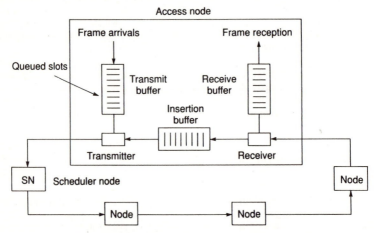

Fig. 10.23 *CRMA-II system configuration*

transmission media to form a closed loop. The figure shows a number of logical functions present in a node for operation of the CRMA-II protocol. In particular, the relative positions of the receive buffer, the insertion buffer and the transmit buffer within the nodal architecture are shown.

A node's receiver function recognises data destined for this node from the destination address carried in the slot. It therefore copies the data into the receive buffer and frees the slot. The free slot is then passed to the insertion buffer. If the insertion buffer is empty, slots entering are output to the transmitter function immediately and no insertion delay is involved.

The transmit buffer holds data in the form of slots destined for transmission onto the ring. Data are queued here to await a transmission opportunity, which requires the reception of a free slot from the insertion buffer (signifying that the medium is idle and the insertion buffer is empty). The node may then begin its transmission; once initiated, a whole frame of data is transmitted as a series of contiguous slots. Any busy slots arriving from upstream that are not destined for this node are queued in the insertion buffer during the node's own transmission. Subsequently, the arrival of free slots will enable the insertion buffer to empty. In effect, a free slot arriving at a non-empty insertion buffer is cancelled to shorten the queue by one slot. Clearly, for correct operation the size of the insertion buffer must be at least equal in slots to the size of the maximum length frame to be transmitted.

One of the nodes on the ring - the so-called scheduler node - is chosen to perform slot monitor and reservation mechanism functions. All nodes will generally be equipped to perform these functions but only one scheduler will be active at any one time.

In operation, slots are transmitted from node to node around the ring. The situation is dynamic, with busy slots being created by source nodes and freed by destination nodes in a random manner. Each node examines the destination address of each busy slot that it receives for a match with its own address. When such a match occurs, the node will copy the slot data into its receive buffer. Providing the destination address was unique to this node (i.e. not multicast), the node will then either

- replace the received busy slot by transmitting a free slot, or

- 'cancel' the received slot in order to shorten the queue in the insertion buffer, or

- replace the slot with one of its own slots queued in the transmit buffer (but only, of course, provided the insertion buffer is empty).

Free gratis slots are available to all nodes for unrestricted access. In light traffic conditions, gratis slots provide fast access, good network throughout and minimal overall delay. In heavier traffic conditions a reservation mechanism creates enough reserved slots to ensure that all nodes receive fair access to the medium and that access delays are tightly bounded.

10.5.5 Slot reservation mechanism

The slot reservation mechanism operates in a cyclic manner under control of the scheduler node. It can be arranged to 'cut-in' only when the traffic load on the system reaches some pre-arranged thresholds

The reservation cycle begins when the scheduler issues a MAC reserve command. This consists of a start delimiter and end delimiter transmitted back-to-back. The reserve command is passed from node to node, with each node inserting two parameters into it. These are the combined number of slots awaiting transmission in the node's transmit and insertion buffers (a_i, say), plus the number of slots transmitted from the node's transmit buffer (A_i, say) since the last reserve command was received.

The reserve command arrives back at the scheduler node containing the a_0, a_1.....a_n and A_0, A_1....A_n values for every active node on the system. The scheduler then sums the two parameters a_i and A_i for each node and, by means of an appropriate algorithm, truncates the result to arrive at a fairness threshold T. By subtracting parameter A_i from threshold T for every node, the scheduler computes the number of reserved slots to be created.

Having performed the scheduling computation, the scheduler issues a confirm command containing the threshold value T. This enables each node to deduce the number of reserved slots allocated to it as $(T-A_i)$. If the value of $(T-A_i)$ happens to be negative, this is taken by the node as an instruction to defer any further transmissions from its transmit buffer until $(T-A_i)$ free gratis slots have passed by.

After sending the confirm command, the scheduler node marks all passing free gratis slots as reserved until the scheduled number of reserved slots have been created. This brings the reservation cycle to an end, ready for the start of the next one.

Fig. 10.24 shows the three phases of the slot reservation mechanism for a 3-node system.

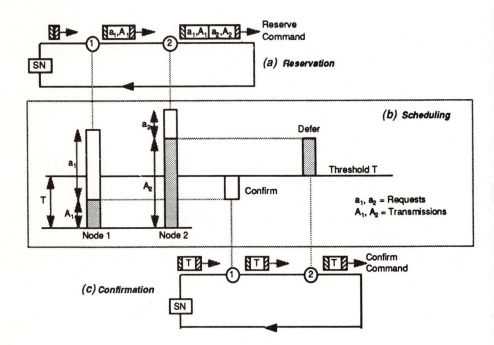

Fig. 10.24 *CRMA-II protocol reservation scenario*

The reservation phase is illustrated at (a). This shows the scheduler node issuing a reserve command which circulates the ring to node 1. Here, the two parameters a_1, A_1 are added (combined number of slots queued in the nodes transmit and insertion buffers, number of slots transmitted from transmit buffer since last reserve command). The lengthened reserve command then continues to node 2, where it is further lengthened by the addition of parameters a_2, A_2. Finally, from node 2 the reserve command arrives back at the scheduler node.

The scheduling phase is illustrated at (b). This shows the actions at the scheduler node, in adding the two parameters received from each of the nodes - (a_1+A_1) and $(a_2 + A_2)$ - and truncating the results at a fairness threshold T. Then, by computing the values $(T-A_1)$ and $(T-A_2)$ the scheduler determines the number of reserved slots to be created. Note that, since the value of $(T-A_2)$ is negative, no reserve slots will be created for node 2.

The final confirmation phase of the cycle is illustrated at (c). This shows the scheduler node issuing the confirm command containing the threshold value T. The command circulates the ring to inform all nodes of the threshold value. From this, node A can deduce that it has been allocated $(T-A_1)$ reserved slots. On the other hand, since $(T-A_2)$ is a negative number, node B deduces that it must defer transmitting from its transmit buffer until $(T-A_2)$ gratis slots have passed.

10.5.6 Immediate access scenario

Immediate access is the opportunity to transmit a frame of data newly arrived in the transmit buffer with negligible delay. This generally occurs at a node that is experiencing light traffic loads, so that an abundance of free gratis slots is available. The node will thus begin its transmission in the first free gratis slot following frame entry into its transmit buffer.

Once frame transmission has started, it continues to completion using contiguous slots. Any busy slots arriving at the node during the transmission time are queued in the node's insertion buffer. However, on completion of the node's frame transmission, the frequency of arrival of free gratis slots will ensure that the insertion buffer is quickly emptied.

On the subsequent arrival of a reserve command, since the node's transmit and insertion buffers are both empty it will make no slot reservations. However, as required by the protocol it reports the number of slot transmissions made since receipt of the last reserve command.

In the event of a node neither having slots queued for transmission nor any previous slot transmissions to report, it makes no entry into the reserve command.

Fig. 10.25 shows three typical phases of an immediate access scenario. In (a), a node is shown transmitting a 4-slot frame of data onto the medium. It has just completed transmission of the second slot and the insertion buffer is currently empty, although a frame of data comprising 4 slots labelled 1 from upstream is about to arrive. Since the transmit buffer still has 2 slots pending, two of the slots arriving from upstream will need to be queued in the insertion buffer.

In (b), the node has completed its frame transmission and its transmit buffer is empty. Two of the busy slots labelled 1 from upstream have passed the node and the other two busy slots are queued in the insertion buffer. However, two consecutive free gratis slots are about to arrive at the node and these will enable the insertion buffer to empty. The reserve command following the two free gratis slots will thus arrive at a node at which no slots are queued for transmission.

In (c), the last of the slots labelled 1 has been transmitted past the node, followed by the reserve command. This has had parameters 0, 4 inserted by the node, denoting that it has no slots pending (so requires no reserved slots) and that it has transmitted 4 slots since the previous reserve command.

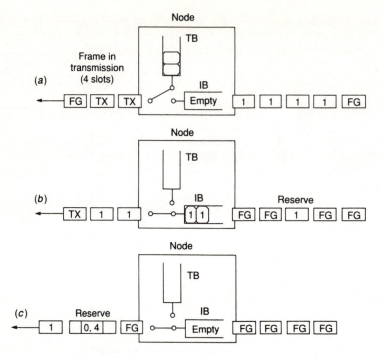

Fig. 10.25 *CRMA-II protocol immediate access scenario*

10.5.7 Reservation access scenario

A node encountering severe congestion, with no free gratis slots passing by, might find itself with a frame of data awaiting transmission in its transmit buffer and busy slots queued for transmission in its insertion buffer. If, under these conditions, a steady stream of busy slots continues to arrive, this will prevent the insertion buffer from emptying.

The arrival of a reserve command enables the node to make a reservation request equal to the sum of the slots queued in the transmit and insertion buffers. It also enables the node to report a zero transmission count.

Subsequently, the confirm command arrives containing the threshold value T. This should indicate that the node has been allocated sufficient reserved slots to enable it to empty its insertion buffer. Even if the congestion persists, the node then merely awaits the arrival of its allocated number of reserved slots which will enable it to transmit its pending frame.

The five diagrams given in Fig. 10.26 show five typical phases of the reservation access scenario.

In (a), the node has a 3-slot frame in its transmit buffer pending transmission and two busy slots queued in its insertion buffer. A steady stream of busy arriving slots will prevent it from emptying its insertion buffer and starting its own frame transmission. However, the imminent arrival of a reserve command will present an opportunity to request some reserved slots to alleviate the congestion.

In (b), the node has transmitted the reserve command onto the medium after inserting the parameters 5, 0. These signify that the node has 5 slots pending and has transmitted zero slots since the previous reserve command. The congestion persists, with busy slots constantly occupying the medium. Note that the reserve command, which arrived at the node behind the first slot labelled 2, leaves the node in front of both that slot and the last slot labelled 1. This is because a MAC command, on arrival at a node, is permitted to bypass the contents of the insertion buffer.

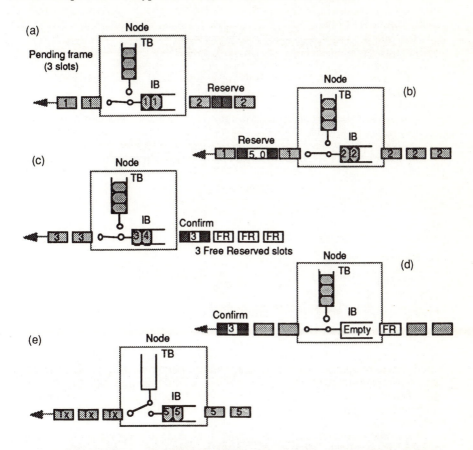

Fig. 10.26 *CRMA-II protocol - reservation access scenario*

In (c), the confirm command from the scheduler, carrying the threshold value of 3, is about to arrive at the node. This will inform the node that (3-x) reserved slots have been created for its use, where x in this instance is zero (the number of slot transmissions from the node in the preceding reservation cycle). Thus, the 3 free reserved slots immediately following the confirm command are available for use by the node.

In (d), the confirm command has been relayed onwards by the node, followed by two busy slots labelled 4. Note that 2 of the 3 reserved slots have been used to empty the insertion buffer. The 3rd free reserved slot is just about to arrive at the node, which will allow the node to transmit the contents of its transmit buffer.

In (e), the 3rd free reserved slot has been used by the node for the first slot of its queued frame. The second and third slots of the frame then followed contiguously, which has meant that two arriving busy slots from upstream have been queued in the insertion buffer.

10.5.8 Deferred access scenario

If a node has a frame of data in its transmit buffer awaiting transmission and a confirm command arrives, and if the value of the threshold count T carried is less than the reported value A_1 of the node's previous transmissions, the node deduces that it must defer transmission until (A_1-T) free gratis slots have passed.

The reservation process thus has two beneficial effects:

– it enables the scheduler node to create reserved slots for the benefit of nodes that are being 'starved' of transmission opportunities

– it enables the scheduler node to throttle back the demands of nodes that are hogging bandwidth.

The five diagrams given in Fig. 10.27 show 5 typical phases of a deferred access scenario.

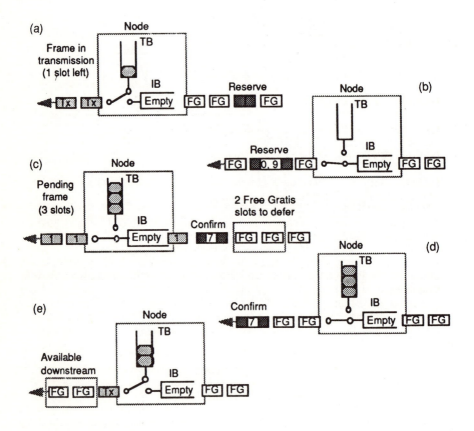

Fig. 10.27 *CRMA-II protocol deferred access scenario*

In (a), the node is about to complete the transmission of a frame of data from its transmit buffer by sending the final slot. Enough free gratis slots are about to arrive at the node to ensure that the insertion buffer remains empty. A reserve command will arrive at the node imminently.

In (b), the reserve command has been transmitted by the node which added parameters 0 and 9, signifying no pending transmissions but a transmission count of 9 since the previous reserve command.

In (c), the node has received a frame of data in its transmit buffer which it has been unable to send because of a passing stream of busy slots. The confirm command is about to arrive at the node; it carries a threshold value of 7, which is less than the node's reported transmission count of 9. The node will therefore deduce that it must defer its pending transmission until two free gratis slots have passed by.

In (d), the confirm command has been relayed forward by the node, which has also allowed two free gratis slots to pass despite having data queued in its transmit buffer.

In (e), the node has started transmitting its pending frame after deferring for the two free gratis slots, since further free gratis slots arrived to present the transmission opportunity.

10.6   Bibliography

1. ISO/IEC 9314-1 : FDDI Physical Layer Protocol

2. ISO/IEC 9314-2 : FDDI Medium Access Control

3. ISO/IEC 9314-3 : FDDI Physical Medium Dependent

4. ISO/IEC DIS9314-4 : FDDI Station Management

5. ISO/IEC 8802.6 : Distributed Queue Dual Bus (DQDB) Subnetwork of a Metropolitan Area Network (MAN)

6. Specification of the Asynchronous Transfer Mode Ring (ATMR) Protocol - submission of the Japanese National Body to ISO/IEC SC6

7. van As, H.R., Lemppenau, W.W., Zafiropulo, P. and Zurfluh, E., 'CRMA-II : A Gbit/s MAC Protocol for Ring and Bus Networks with Immediate Access Capability'.

Chapter 11

Data services over cellular radio

Ian Harris

11.1   Introduction

A cellular radio network comprises a network of radio cells in which each adjacent cell operates on a different radio frequency (RF) carrier. Cells which are not adjacent to each other may use the same carrier frequency. Each cell is prevented from interfering with another cell operating on the same carrier frequency by limiting its power output or radiation pattern. A seven cell repeat pattern is typical of a cellular network architecture whereby a central cell surrounded by six other cells may be repeated and requires only seven RF carriers for widespread coverage. Each cell has independent transmit and receive frequencies which enable simultaneous both-way (full duplex) communications, which has enabled the full use of the Public Switched Telephone Network (PSTN) to be extended to the mobile environment.

Although PSTN interworking with certain private mobile radio (PMR) systems have existed for some years, their growth and success has been limited for a variety of technical and commercial reasons; e.g. application-specific products and the inability to support full duplex communications. The cellular radio telephone network offers a much more open architecture and does not suffer from many of the limitations of PMR systems.

There are two established UK cellular radio telephone operators - Cellnet and Vodafone. Both operators offer virtual nationwide coverage on an analogue cellular telephone network known as TACS (Total Access Communications System) which is based on the American standard known as AMPS (Advanced Mobile Phone System). Both operators also have licences to operate a digital cellular telephone network known as GSM (Groupe Special Mobile). GSM is a Pan-European network conforming to standards produced under ETSI (European National Standards Institute). GSM is at various stages of deployment throughout Europe including the United Kingdom.

Whilst the traffic carried on a cellular radio telephone network is predominantly voice, it is also used for data communications whereby the mobile station is able to interwork with data equipment in the PSTN and data-only networks such as British Telecom's PSS (GNS).

The term 'data communications' is used here in the broadest sense and also embraces such equipment as facsimile.

It is estimated that the number of conventional data users (excluding facsimile etc) in the UK TACS cellular telephone network is between 1% and 2% of the total number of mobile subscribers. Some bolder estimates put the projected figure for the GSM network at 5% for the same type of services.

The figure by all accounts is small, and in the UK TACS network there are very few data-only users. Most require the capability of speech as well as data. In situations where speech is not a requirement then, depending upon the application, data-specific radio networks such as 'Paknet' (a public cellular packet radio network) are often able to offer the user a more cost-effective solution. At present there seems to be little indication that GSM will change this situation because of technology costs and tariffs.

11.2   Data in the UK (TACS) analogue cellular network

The TACS network operates in the frequency band 890 MHz to 950 MHz with a channel spacing of 25 kHz. Each channel is full duplex and its performance is similar to that of a 3.1 kHz audio telephone circuit.

Both UK cellular operators support data communications but in different ways. Vodafone's approach is to use cellular specific modems and data gateways in order to provide interworking with the fixed network environments, thereby obviating the need for the mobile user to have any prior knowledge of the fixed network environment or its equipment.

Cellnet favours more strongly the use of conventional PSTN modems (which enable direct interworking with a PSTN modem of the same type) but is also supportive of the need for cellular-specific modems, thereby offering the user a greater freedom of choice.

The low usage of 'data' in a cellular environment and the inability of the two operators to agree on a common approach has done little to encourage mobile phone manufacturers to make provision for data in their phones or even to agree on a standard interface.

There are however some fundamental differences between the cellular radio environment and that of the PSTN. These differences are agreed upon by both operators to a greater or lesser degree and are discussed below.

11.2.1 Cellular radio transmission impairments

Radio transmission impairments in the cellular radio environment have no equivalent in a conventional PSTN environment or any fixed network environment.

Obstacles, such as buildings or vehicles, in the radio path can put the receiver in a radio shadow or cause the radio signals to be reflected such that a cancellation of signal levels occur at the receiver. The former impairment, known as 'shadowing', and the latter, known as 'multipath', are illustrated in Fig.11.1.

A received radio signal has an inherent noise level at the receiver. This noise may be due to the radio environment, or noise within the receiver circuitry itself. A typical FM radio receiver may have a noise level of about -120 dBm. The wanted received signal level in a cellular environment may be as low as -110 dBm.

The effect of multipath and shadowing is to cause 'fading' whereby the wanted signal momentarily falls below the noise threshold. In severe conditions the effect may be prolonged and after approximately 5 seconds the network or the cellular phone will terminate the call. If the obstacle, the transmitter or the receiver are moving, the problem is compounded. Fig. 11.2 shows the effect of fading and shadowing on a wanted radio signal.

Two further problems arise in cellular communications. The first is 'power stepping', which is a technique used by cellular radio base stations to change the power level of mobiles as they approach or move away from a base station site. The second is 'handover', which occurs when a mobile moves from one cell boundary to another.

Data services over cellular radio

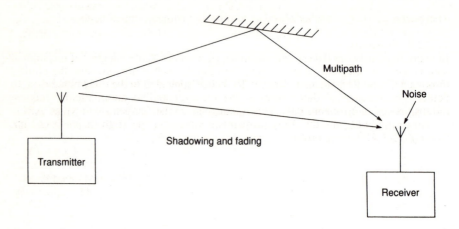

Fig. 11.1 *Radio channel impairments*

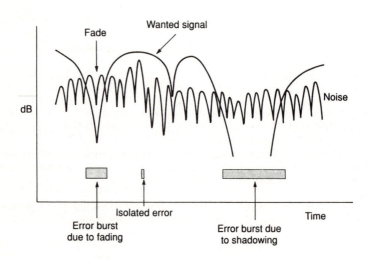

Fig. 11.2 *Effect of fading and shadowing on wanted signal*

Both have the effect of momentarily disrupting the radio carrier for a few hundred milliseconds.

It is fairly common within a cell boundary for the receiver to experience bit error rates of the order of 2% (i.e. 1 in 50 bits may be in error). By contrast, PSTN bit error rates are rarely worse than 1 in 10^5.

11.2.2 Error protection methods

There are a number of methods used to combat transmission impairments.

11.2.2.1 Protocols

In many fixed network data communications environments, the integrity of transmitted data is usually safeguarded by a layer 2 protocol. Such protocols have the ability, through a Cyclic Redundancy Check (CRC) code appended to the transmitted data, to detect the presence of data errors at the receiver and to automatically request retransmission of erroneous data. This retransmission mechanism is known as 'ARQ'.

An example of how the ARQ mechanism works for the High Level Data Link Control (HDLC) layer 2 protocol standard is shown in Fig.11.3.

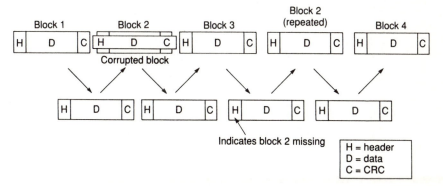

Fig. 11.3 *High Level Data Link Control (HDLC)*

The data to be transmitted is divided up into blocks, each block having a CRC checksum. The header of each block contains supervisory information and, in particular, a send and receive block sequence number, which indicates the number of the last sent and last received block within a modulus window.

If an erroneous block is received, the receiver informs the transmitter which then retransmits the block at the earliest opportunity.

In severe error conditions, it is possible for data transmission to cease through an effect known as 'windowing'. Windowing will occur when the outstanding erroneous blocks are unable to be successfully transmitted within the block sequence number modulus window.

To improve the performance of a layer 2 protocol in a noise-prone environment a technique known as Forward Error Correction (FEC) may be used.

11.2.2.2 Forward Error Correction

The basic principle of Forward Error Correction (FEC) is to add redundant information to the source data so that, in the event of bit errors occurring during transmission, the receiving entity is able to use the redundant information to correct the bit errors without the need for ARQ. Clearly there is a limit to the number of bit errors which may be corrected in this way and so, in practice, FEC merely reduces the frequency with which ARQ occurs.

There are a number of different FEC codes, e.g. symbol block codes, convolutional codes and binary block codes. The Reed Solomon (RS) codes are probably the best known symbol block codes and tend to be more effective in dealing with burst errors rather than random errors. Convolutional codes tend to be more complex to implement. Binary block codes can be very effective in correcting for a large number of random bit errors and are relatively simple to implement. Probably the best known binary block codes are the BCH (Bose-Chaudhuri-Hocquenghem) codes. A compromise between implementation complexity, code performance in the given environment and redundancy overhead often has to be made when selecting an FEC code.

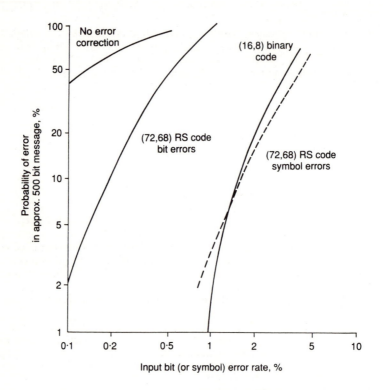

Fig. 11.4 *Performance of RS and BCH codes*

Fig. 11.4 shows the relative effectiveness of an RS code and a BCH code. The degree of redundancy is expressed as two numbers. In the case of the RS example there are a total of 72 bits in each code block of which 68 are the source data. In the case of the BCH code, there are 16 bits in each code block of which 8 are the source data. The solid curves show the performance when the errors are randomly distributed. The dotted curve shows the performance of the RS code when the errors occur in bursts. The performance of the BCH code can be poor under burst error conditions. In order to strengthen the effectiveness of such a code, a technique known as 'interleaving' is often used.

11.2.2.3 Interleaving
Once FEC has been applied to the source data, the data stream may be considered as being comprised of a number of logical blocks called code blocks. To apply interleaving, the bits of each code block are spread out in time by interleaving them with bits of other code blocks. The new distance between any two successive bits belonging to the same original code block is called the 'interleaving depth'.

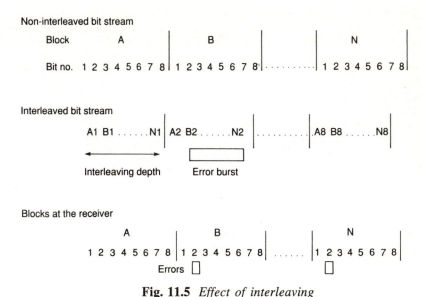

Fig. 11.5 *Effect of interleaving*

Fig. 11.5 shows the effect of an error burst occurring in an FEC encoded non-interleaved bit stream compared to the effect of a similar burst occurring in an FEC encoded interleaved bit stream. Let us assume that the FEC algorithm is capable of correcting up to 2 bits in error in any code block. In the non-interleaved case, if the error burst spreads over, say, 4 bits then the FEC algorithm would be unable to correct the error. In the interleaved case, the same error burst will only affect 1 bit in each original code block and so, after de-interleaving, the FEC algorithm would be able to correct the error.

The limitation of the effectiveness of interleaving is of course when the error burst is greater than the interleaving depth. Increasing the interleaving depth has the disadvantage of increasing the transmission delay.

Modem Type	Bit Rate (bit/s)	Modulation	Equalisation	Scrambler
V21	300/300	PSK	No	No
V23	1200/75	PSK	No	No
V26bis	2400/150	QPSK/PSK	Fixed	Optional
V22	1200/1200	QPSK	Fixed	Yes
V22bis	2400/2400	16pt APK	Adaptive	Yes
V22ter	4800/75	8 phase PSK/PSK	Adaptive	Yes
V32	9600/9600	16/32 phase APK (Trellis)	Adaptive	Yes

Fig. 11.6 *Modem characteristics*

11.2.3 *PSTN modem performance in a cellular environment*

A conventional PSTN modem may be attached to a number of cellular phones by way of an adaptor box specially designed for a particular mobile phone.

Fig. 11.6 lists some of the characteristics of some commonly used modems. V22bis and V32 modems invariably offer V42 error correction. V42 is basically a layer 2 ARQ protocol.

The performance of a PSTN modem (with or without V42) in a cellular environment tends to be unpredictable, primarily because such modems are intolerant of high bit error rates and sustained carrier loss. V22bis, V29 and V32 modems also have complex synchronisation methods at the start of a call which do not perform well in high bit error rate conditions.

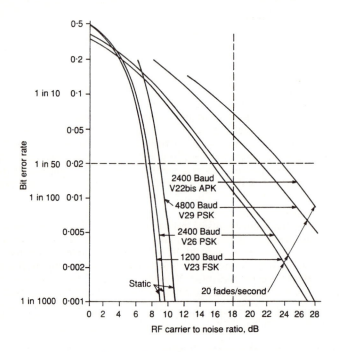

Fig. 11.7 *Modem performance*

Fig. 11.7 shows the measured performance of V23, V22bis, V26 and V29 modems in a static and a mobile environment. 20 fades per second corresponds to a vehicle movement of about 20 kmph. The 18 dB reference corresponds closely with a cell boundary design threshold; i.e. to give a short term mean Carrier-to-Noise (C/N) and Carrier-to-Interference (C/I) ratio in excess of 18dB over 90% of the cell area.

It can be seen that, for 1 in 50 (2%) bit error rate and an 18dB threshold, neither the V29 modem at 4800 Baud or the V22bis modem give adequate performance. V29 and V22bis modems also include scramblers to improve the performance of the demodulator. Polynomial de-scramblers have a tendency to propagate errors (error multiplication) when receiving data from a high bit error rate environment.

V29 and V22bis modems require a wider bandwidth and are thus more affected by amplitude and phase distortion incurred in the audio processing stages of a cellular transceiver.

Some modems have adaptive equalisers which malfunction under fading conditions. Their performance worsens by some 2dB when equalisation is included, primarily because of attempts by the equaliser to try to adapt to an unstable line condition.

V21 and V23 modems can only provide low data rates which are not generally acceptable to users.

The performance of V26bis at 2400 bit/s offers the best performance in a cellular radio environment.

It is reasonable to conclude that conventional PSTN modems can appear to work reasonably well under certain conditions (e.g. when static). Their performance, however, is somewhat unpredictable particularly when mobile and when subjected to the transmission impairments described earlier.

11.2.4 Cellular-specific modems

The most common cellular-specific modem in use in the TACS network is the 'CDLC modem'. The CDLC modem incorporates a layer 2 forward error corrected and interleaved protocol known as CDLC (Cellular Data Link Control). CDLC is in the 'public domain' and was developed in 1985 by Racal Vodafone Ltd (now Vodafone Ltd) in conjunction with Racal Research Laboratories.

The modem itself is manufactured by third parties and interfaces to different cellular phones using an appropriate interface cable and phone-specific software.

The physical interface is a 25-way 'D type' connector allowing the attachment of any asynchronous terminal conforming to CCITT V24/V28 standards. The modem also incorporates auto calling/auto answering procedures conforming to CCITT V25bis.

11.2.4.1 CDLC modem modulation
The modem performance curves discussed earlier indicate that the best modem choice would be V26bis. Such a modem is normally only used in its full-duplex mode on leased lines and so, in order to allow the V26bis modem to interwork with the PSTN, a data gateways is required. However, in some cellular environments abroad, where data gateways might not be possible, the performance of the backward channel of V26bis was assessed at 150 bit/s. Its performance over the cellular radio path indicated that it was adequate to enable the CDLC modem to interwork directly with the PSTN in an asymmetric full-duplex mode with certain limitations which are discussed later.

11.2.4.2 Cellular Data Link Control protocol
CDLC is a layer 2 protocol based on IS4335 (HDLC) standards. The elements and procedures of HDLC have been retained, including selective reject for more efficient error recovery. Frame synchronisation has been changed since the normal HDLC flags (7E Hex) are intolerant of high bit error rates.

CDLC layer 2 procedures overlay BCH and Reed Solomon FEC codes and interleaving.

Fig. 11.8 shows the comparison between the HDLC and CDLC frame structures. The 48-bit CDLC leading synchronisation pattern has inherent error protection and is not included in the CRC checksum. Neither is it subjected to the FEC and interleaving algorithms which apply to the rest of the frame. CDLC has eight different 48-bit synchronisation codes. Each code defines one of eight frame lengths, thereby obviating the need for a trailing flag. The reason for the different frame lengths is to minimise the transmission delay when the input data is less than that which can be carried in the maximum frame size.

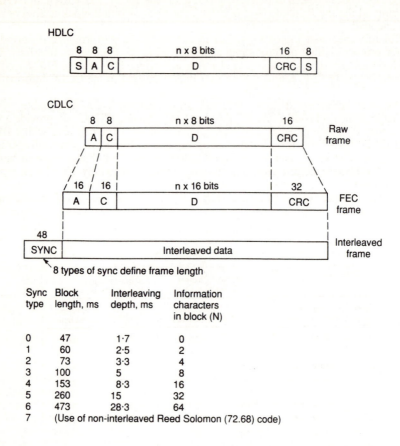

Fig. 11.8 *Comparison of HDLC and CDLC frame structures*

The synchronisation codes have a low probability that they will occur in a random data stream and a high probability that successful synchronisation will be achieved in the event of errors. Each synchronisation code will tolerate up to 8 bits in error in any position. The probability of a random data stream being decoded as one of the eight synchronisation codes is 9.1×10^{-6}. The probability of successful synchronisation with a random bit error rate of 2% is 99.96%

The FEC codes used in CDLC are the 16,8 BCH code and the 72,68 RS code. The code chosen for the backward channel mentioned earlier was a 23,12 Golay code.

The redundancy overhead in the BCH code restricts the maximum user rate which can be supported to just over 1200 bit/s. The RS code redundancy overhead is much lower and allows a user rate in excess of 2400 bit/s to be supported. However, the RS code is weaker than the BCH code (particularly for random bit errors) and so the CDLC modem incorporates an automated strategy for selecting when a particular FEC code is to be used. The RS code is selected when the bit error rates are low. When the bit error rate increases, the BCH code is selected and remains in use until the bit error rate decreases. A number of factors are taken into account before the decision to change FEC codes is taken as it is essential to avoid instability. In the event of an ARQ occurring whilst the RS code is in use, then the re-transmitted block is always sent BCH encoded.

Probably the most important factor in assessing the performance of any data transfer protocol is that of throughput as perceived at the application interface. The asynchronous user data start and stop elements are stripped off prior to FEC encoding and therefore only FEC encoded 8-bit user data is carried in the CDLC information field.

The cell boundary design threshold signal strength for acceptable voice communications is -110dBm. At this threshold, RS encoded CDLC will provide a full duplex user data rate of 2430 bit/s over typically 80% of the cell area whilst BCH encoded CDLC will provide a full duplex user rate of 1340 bit/s over typically 90% of the cell area. Therefore, user data rates of 1200 bit/s and 2400 bit/s can be sustained even under adverse radio conditions allowing for approximately 10% block repeat through ARQ.

The performance of the BCH and Golay FEC codes used in CDLC are shown in Fig. 11.9. An ARQ will occur whenever the percentage of block errors is greater than zero.

Physical Conditions		Percentage Block errors (after FEC)			
		2400 baud Channel 512 data bits (BCH)		150 baud Channel 36 data bits (Golay)	
Speed km/hr	Signal Strength dBm	No FEC	Interleaved 16, 8 code	No FEC	Interleaved 23, 12 code
0.20	-120 to -110	23.7	0.6	8.8	0.0
0-20	-110 to -100	17.5	0.6	5.3	0.0
0-20	-100 to -90	11.1	1.0	5.4	0.0
0-20	-90 to -80	8.3	0.0	4.5	0.0
20-40	-120 to -110	49.8	0.0	24.2	0.0
20-40	-110 to -100	42.6	0.0	23.1	0.0
20-40	-100 to -90	27.8	0.2	11.7	0.0
20-40	-90 to -80	18.1	0.0	4.4	0.0
40-75	-120 to -110	58.5	1.7	---	---
40-75	-110 to -100	36.1	0.0	---	---
40-75	-100 to -90	29.2	0.6	4.2	4.2
40-75	-90 to -80	'17.8	0.0	2.0	2.0

Fig. 11.9 *Performance of BCH and Golay codes*

11.2.5 Interworking between cellular and fixed networks

A PSTN modem attatched to a cellular phone is able to directly interwork with a modem in the fixed network provided they are of compatible types. For specific and dedicated applications in a closed operational environment, this is often satisfactory, other than for the performance problems described earlier. However, for the more open applications, it is not always possible for a mobile user to have prior knowledge of the type of modem being used by the 'called party' ('b' party) in the PSTN. In order to overcome these difficulties, a data gateway is necessary which standardises the mobile user's access and isolates the mobile user from the multiplicity of standards in use in the fixed networks.

The majority of CDLC modem users access the fixed network through such a data gateway provided in the Mobile Switching Centre (MSC). The data gateway (shown in Fig. 11.10) provides CDLC modem interworking with the following fixed networks:

PSTN - V21, V22, V23, V22bis and V42 error correction
PSPDN - PSS (GNS), Mercury, IBM, ISTEL
PRIVATE NETWORKS

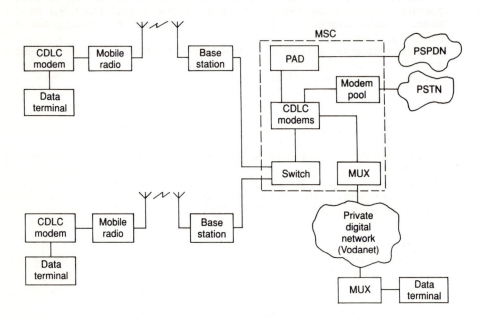

Fig. 11.10 *Data gateway*

11.2.5.1 PSTN interworking via the data gateway

To access any modem in the PSTN, the mobile user prefixes the PSTN STD code and 'b' party number with 972, which identifies the call at the MSC as a PSTN data call so that the appropriate resource can be reserved in the data gateway. The STD code and 'b' party number is routed into the PSTN (without the need for two-stage call set up) as in a normal speech call. When the 'b' party answers, the reserved PSTN modem in the data gateway automatically configures itself to be compatible with the characteristics indicated by the answering 'b' party modem. At the same time, the CDLC link level is established between the mobile CDLC modem and the CDLC modem in the MSC.

Out-of-band flow control using the V24 control signal clear-to-send (CTS) is provided by the data gateway so that the service is transparent to all characters in the CCITT IA5 character set. Flow control is essential because the data rate used by the mobile DTE may be different to that used by the PSTN modems or by the DTE attatched to the modem in the PSTN.

The service also provides interworking with private analogue (leased line type) circuits, which is commercially attractive to large corporate users. In such a case, a private access code is used instead of an STD code when making the call.

11.2.5.2 *Packet switched network interworking via the data gateway*

To access an X25 packet switched network, the mobile user dials 970xxx, which connects the mobile to a PAD (Packet Assembler/Disassembler) on the appropriate network defined by the digits 'xxx'. Once connected to the PAD, conventional X3/X28 PAD control procedures and X121 call establishment procedures for X25 networks apply.

The PAD belongs to and is maintained by the appropriate packet switched network operator. In order for the mobile user to access any of the packet switched network hosts, the mobile user usually has to own a password issued by the packet switched network operator.

This service does not allow a host in the fixed network to call a mobile. This is not only due to a technical limitation of the PADS themselves, but also due to commercial complications concerning 'billing'.

11.2.5.3 *Private digital networks via the data gateway*

The data gateway provides the capability to interwork with private digital networks. As is the case with analogue leased lines, this service is attractive to large corporates who are able to justify the higher capital cost outlay against the low-cost call tariff.

To access the service, a mobile user prefixes a private network access code with 973. This service uses digital multiplexers instead of modems. The interconnection to the fixed network is usually a 64 kbit/s digital channel which may be one of 30 such channels carried in a 2 Mbit/s 'megastream' connection. Up to 24 simultaneous data calls can be in progress on one 64 kbit/s channel.

The 973 service also allows the host in the fixed network to call a mobile user.

11.2.5.4 *Test service (internetworking with PSTN)*

This test service allows a user in the PSTN having a conventional PSTN modem to interwork with a mobile having a CDLC modem.

When a cellular phone is dialled from the PSTN, there is no simple way for an MTX to discriminate between a data call and a voice call.

If a user in the PSTN wishes to establish a data call with a mobile he must first dial 0836 331441.

When access to the 331441 service is confirmed, the PSTN calling party is then required to enter the mobile number required. On successful completion of the second stage of the call, the data connection is established.

The 331441 service provides modem interworking capabilities with the PSTN identical to those provided by the 972 access service described earlier.

11.2.5.5 *Test service (operational checking)*

A 970123 test service number is provided to allow mobile users to check the operation of their mobile data equipment. The test service has a number of facilities which can be accessed by appending appropriate digits to the end of 970123 when dialling. For example, 97012311 will provide a menu of facilities offered such as a 60 second transmission of scrolling CCITT IA5 printable characters with a 'beep' every 1200 characters. An automated dialback data service is also provided.

11.2.5.6 *Mobile to mobile CDLC calls*

Mobile to mobile calls do not go via the data gateway and operate in the symmetric full-duplex mode, i.e. 2400 bit/s in both directions simultaneously.

11.2.5.7 *Direct interworking with the PSTN*

As mentioned earlier, the CDLC modem is capable of interworking directly with the PSTN without going via the data gateway. This mode of operation is used very little in the UK.

In order to interwork directly with the PSTN, the user must select the 'asymmetric full duplex mode' (sometimes referred to in CDLC as the '2-wire' mode) in the CDLC modem. The operation of the layer 2 CDLC protocol itself is unchanged; that is, the protocol performs as if the symmetric full duplex mode (2400 bit/s both directions) were being used. In the asymmetric full duplex mode, the forward channel operates at 2400 bit/s and the backward channel at 150 bit/s. The main limitation is that applications requiring mass data transmission simultaneously in both directions are restricted in one direction to 150 bit/s. Most applications are half duplex by nature and so the performance is virtually identical to that of the symmetric full duplex mode. The 2400 bit/s direction is automatically selected according to the direction in which most data is flowing. If the application subsequently requires mass data to flow in the opposite direction then a line reversal procedure takes place automatically. The line reversal procedure is protected against inadvertent triggering and will only be invoked if more than two or three characters are queued awaiting transmission, even if there is no data flowing in the opposite direction. Some applications, such as 'Videotex', are therefore possible with virtually no visible degradation in performance.

11.2.6 *Facsimile*

There are literally millions of facsimile machines in use throughout the world, the majority of which are group 3 machines conforming to the CCITT T30 protocol. The encoding of information contained on facsimile pages conforms to CCITT T4.

It is probably necessary to understand a little about how a facsimile machine works in order to appreciate its operational limitations in a cellular environment.

The CCITT T30 protocol has a control phase and a message phase. The control phase operates at 300 bit/s half duplex and is responsible for negotiating the capabilities between facsimile machines and supervising the transmission of pages of information. An ARQ mechanism is built into the control phase procedures.

The message phase operates at 9600, 7200, 4800 or 2400 bit/s half duplex and is responsible for transmitting pages of facsimile information encoded according to CCITT T4. A T4 encoded page of A4 text may typically comprise 40,000 bits.

At the start of a facsimile call, the answering facsimile machine offers a set of attributes, such as transmission speed, from which the calling facsimile machine must select a compatible set or subset. The calling facsimile machine then sends a modem training sequence (according to the appropriate modulation standard defined in the CCITT modem 'V' series recommendations) followed by a sequence of binary zeros at message phase speed known as the Training Check Flag (TCF).

The purpose of the TCF is to enable the receiving facsimile machine to check the quality of the connection. The receiving facsimile machine indicates the acceptability by sending a control phase response. If the receiving facsimile machine finds the TCF acceptable, then the sending facsimile machine sends the T4 encoded page of information at message phase speed. If the receiving facsimile machine finds the TCF unacceptable, then the sending facsimile machine will send a new modem training and TCF sequence at the next lowest message phase speed.

Any bit errors which occur as a result of noise during the message phase usually appear as extended or missing lines on the received facsimile image, sometimes accompanied by black lines.

CCITT T30 provides an optional error correction mode. If both the sending and receiving facsimile machines declare and select the error correction mode during the negotiation at the start of the call, then each page of information to be transmitted is split up into partial pages. Each partial page is sub-divided into a number of frames (usually 256) and each frame has its own CRC checksum. In the event of an error being detected in any one or number of frames, the sending facsimile machine is requested to re-transmit those frames which were in error. The error correction process can repeat itself a number of times, thereby reducing the number of outstanding frames that are in error on each repeat, before proceeding to the next partial page.

The CCITT T30 protocol and the optional error correction mode were designed to cope with the bit error rates likely to be encountered on the PSTN. The performance of such a protocol in a cellular radio environment can often be erratic because of the high bit error rates which can occur from time to time.

11.2.6.1 Facsimile in the UK TACS network

Facsimile machines may be attached to a cellular phone using the same adaptor boxes which are used to connect conventional PSTN modems. Unfortunately, facsimile machines contain modems which are intolerant to sustained carrier loss of more than a second or so during the message phase and this often results in premature call termination. The modems operate in the half-duplex mode, which makes them unpredictable as they are unable to discriminate between a line reversal and loss of carrier due to fading. Manufacturers of portable facsimile machines intended for the mobile environment tend to provide extended carrier loss detection of several seconds, but this is only helpful if the facsimile machine with the extended carrier loss is receiving the image, not sending it.

Momentary fading can introduce bit errors into the TCF which will cause most facsimile machines to attempt to re-train immediately at a lower modem speed. Some facsimile machines can retry the TCF at the same modem speed before changing to a lower speed.

There are a number of measures possible which tend to improve the performance of facsimile machines in a cellular environment:

(a)   The facsimile optional error correction mode should be used wherever possible. Unfortunately, both the sending and receiving machine must be capable of operating in the error correction mode and this cannot be guaranteed.

(b)   A significant improvement in performance can be obtained if the facsimile machine is used when the phone is not actually moving.

(c)   Facsimile machines with extended carrier loss timeout in the message phase are less prone to premature call termination.

(d)   Facsimile machines that can make a second TCF attempt at the same modem speed can avoid unnecessary training down to a lower speed.

The use of 'facsimile gateway services' enables data to be transferred over the cellular radio path using facsimile image file transfer protocols developed primarily to transfer facsimile documents between PCs. Such protocols can obtain additional protection through the use of CDLC. However, such techniques preclude the use of conventional facsimile machines in the mobile. Furthermore, the data gateway is unable to allow 'real time' interworking between the mobile and the PSTN facsimile machine since the gateway is essentially a 'store and forward service'.

Facsimile users in the cellular network seem to be content with the reliability they are achieving through direct interworking. This is probably due to the fact that most applications are in fact static.

11.2.7 References and further reading

1. Frazer, E.L., Harris, I. and Munday, P.J., 'CDLC-a data transmission standard for cellular radio'. *Journal IERE,* 1987, **57** (3)
2. CCITT Recommendations V24,V28,V25bis,X25,X28,T4,T30
3. IS4335: International Standards Organisation
4. Bleazard, G.B., 'NCC Handbook of data communications'. NCC publications.
5. Jennings, F., 'Practical Data Communications-Modems, Networks and Protocols'. Blackwell Scientific Publications.
6. Davie, M.C. and Smith, J.B., 'A Cellular Packet Radio Network'. *Electronic Communications Journal,* 1991, (June)

11.3   Data in the GSM digital cellular network

The GSM network operates in the frequency band 890 to 960 MHz with a channel spacing of 200 kHz. Up to 8 or 16 simultaneous traffic channels may be contained within a single radio frequency carrier through the use of Time Division Multiple Access (TDMA). Where 8 such traffic channels are provided each channel is known as a Full Rate Channel (FRC). Where 16 such traffic channels are provided each channel is known as a Half Rate Channel (HRC).

At present only one channel is used for any one purpose at a time, although there may of course be a mixture of speech, data and telematic services in each radio carrier.

The half rate channel has been developed to conserve bandwidth for services which do not require the capacity of a full rate channel.

GSM supports data services in a number of different ways and, in so doing, provides the user with a choice of data rates, error protection method, terminal types, fixed network interworking etc.

GSM also supports telematic and supplementary services. Telematic services invariably have some end to end significance whereby the creation and presentation of user information is standardised either within GSM or by apparatus with specific functionality, for example, facsimile. The GSM short message service is an example of a supplementary service.

GSM architecture is based heavily upon that of ISDN. Unlike the TACS network, there is no necessity for modems in GSM other than those required to interwork with the PSTN.

An Inter-Working Function (IWF) is provided to enable interconnection and interworking with fixed networks.

The transmission impairments affecting data transmission in a GSM environment are very similar to those described earlier for the TACS environment and so there is a similarity in the way the problems are resolved.

11.3.1 Overview of data in the GSM environment

Fig. 11.11 gives an overview of how data is supported in the GSM environment. A data adaptor in the mobile station converts the digital user data rates provided by conventional DTEs to 'ISDN' like rates of 3.6, 6 or 12 kbit/s.

FEC and interleaving are performed on each of these rates to give either 22.8 kbit/s for a full rate channel or 11.4 kbit/s for a half rate channel. The inverse functionality takes place at the base station/MSC except that, in general, the data rate between the base station and the MSC will be either 16 kbit/s or 64 kbit/s depending upon the type of connection between them.

Data rate conversion in the data adaptor and at various stages in the network is achieved through the use of CCITT V110 rate adaptors.

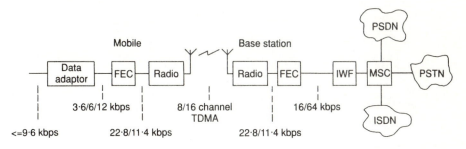

Fig. 11.11 *Data in the GSM environment*

The RA0, RA1 and RA2 V110 rate adaptors are the same as those used in ISDN. The RA1' rate adaptor has been specifically developed for GSM.

The RA0 function converts the asynchronous DTE data rates to comparable synchronous data rates by discarding start and stop elements.

The RA1 function converts the synchronous data rates from the RA0 function into ISDN intermediate rates of 8 or 16 kbit/s.

The RA2 function converts the intermediate rates from the RA1 function into 64 kbit/s.

The RA1' function converts the synchronous data rates from the RA0 function into 3.6, 6 or 12 kbit/s; these being the rates required by the FEC and interleaving function.

The RA1' function produces 60 bit frames, rather than the 80 bit frames used in ISDN, by discarding the synchronisation bits and some of the E (data rate signalling) bits, neither of which are necessary within the GSM environment as frame synchronisation and data rate signalling are achieved by other means.

The FEC and interleaving function is necessary to provide transmitted data with some protection against the high bit error rates which may be present on the radio path.

User Data Rate	Channel	Convolutional Code Redundancy	Interleaving Block Length	Residual BER (after FEC & interleaving)
9.6 kbit/s	Full Rate	61/114=0.53	95 ms	$10^{-2}/10^{-3}$
4.8 kbit/s	Full Rate	1/3	95 ms	$10^{-4}/10^{-5}$
4.8 kbit/s	Half Rate	61/114	190 ms	$10^{-2}/10^{-3}$
<2.4 kbit/s	Full Rate	1/6	40 ms	$>10^{-6}$
<2.4 kbit/s	Half Rate	1/3	190 ms	$10^{-4}/10^{-5}$

Fig. 11.12 *Characteristics of FEC codes*

Fig. 11.12 shows the characteristics of the FEC codes used in GSM and their predicted performance within a cell boundary (i.e. over 90% of the cell area for 90% of the time).

The service provided by GSM with this predicted performance is known as the transparent bearer service and supports the transfer of both synchronous and asynchronous user data. The transparent bearer service has a fixed end to end delay of several hundreds of milliseconds and so any application or layer 2 protocol protection provided by the user must take this into account.

GSM provides an integral layer 2 protocol known as the Radio Link Protocol (RLP). This service is known as the non-transparent bearer service. This service has a variable end to end delay which will always be in excess of that for the transparent bearer service because of the ARQ mechanism used by RLP.

It is likely that the non-transparent bearer service will be used to support a majority of data applications using asynchronous terminals.

11.3.2 The GSM mobile telephone

A mobile telephone may be one of three types as shown in Fig. 11.13.

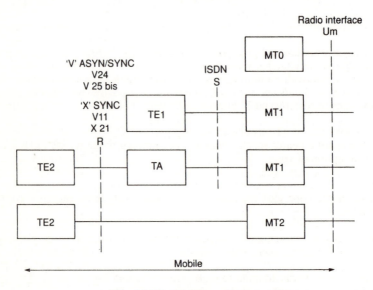

Fig. 11.13 *Mobile telephone types*

An MT0 is a fully integrated mobile station with no external interfaces other than possibly a man-machine interface. Such a station may be a PC with a built-in cellular telephone.

An MT1 provides an 'S' interface to allow the attachment of ISDN computer terminals known as TE1s

An MT2 provides an 'R' interface to allow the attachment of 'V' (asynchronous or synchronous) or 'X' series terminals known as TE2s. Call establishment procedures in accordance with CCITT V25 bis or X21 are also provided using V24 or V11 interface circuits respectively.

In the case where an 'R' interface only is provided, the data adaptor only needs to implement the RA0 and RA1' functions as shown in Fig. 11.14.

11.3.3 The Radio Link Protocol

The Radio Link Protocol (RLP) provides additional protection for user data over and above that provided by the underlying FEC and interleaving.

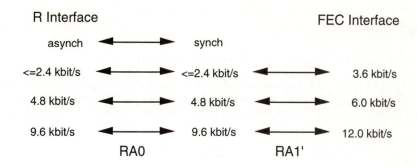

Fig. 11.14 *R interface*

RLP is based on IS4335 balanced class of procedures, more commonly known as HDLC.

The general structure and procedural elements of IS4335 have been preserved in RLP, although some changes were necessary to ensure efficient utilisation of the available radio bandwidth and to ensure that, when the RLP overheads were added with user data at 4.8 kbit/s and 9.6 kbit/s, the required capacity would not be in excess of the FEC rates of 6 kbit/s and 12 kbit/s respectively. The changes were as follows:-

(a) A window size of 64 was chosen to help prevent RLP from 'windowing' due to the transparent bearer service delays. The standard window size in HDLC is 8 with an option of 16.

(b) An FCS of 24 bits was chosen because the HDLC FCS of 16 bits was inadequate to cope with the high bit error rates likely to be encountered on the radio path.

(c) 'Piggy-backing' was adopted to allow information and supervisory frames to be combined in order to make more efficient use of the available radio bandwidth.

(d) Selective reject was adopted to make more efficient use of the available radio bandwidth during error recovery.

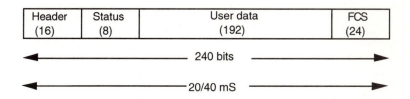

Fig. 11.15 *RLP frame*

An RLP frame (see Fig. 11.15) comprises four 60 bit V110 (RA1') frames giving 240 bits. Each RLP frame is aligned with a radio transmission every 20 ms for a full rate channel and every 40 ms for a half rate channel.

The 240 bits of the RLP frame comprise a 16 bit RLP header, an 8 bit V24 status field, a 192 bit user data field and a 24 bit FCS.

The maximum user data rate which can be supported on a full rate channel and half rate channel given the above constraint is 9.6 kbit/s and 4.8 kbit/s respectively.

11.3.4 Bearer Capability Information Elements

When a call set up is requested from or to a GSM mobile telephone, certain resources within the GSM network have to be assigned for the duration of the call according to the nature of the call. The resources required are requested by the calling entity and conveyed in Bearer Capability Information Elements (BCIE) as part of the signalling requirements at the start of the call.

Typical examples of the content of the BCIE related to data calls are data rate, parity, half or full rate channel. Some of the BCIE options are shown in Fig. 11.16.

8	7	6	5	4	3	2	1	
0	0	0	0	0	0	0	0	octet 1
— Bearer capability IEI								
Length of the bearer capability contents								octet 2
0/1 ext	radio channel requirement		coding std	trans-fer mode	information transfer capability			octet 3
coding standard extension								octet 3a*
1 ext	0 spare	structure		dupl. mode	confi-gur.	0 spare	esta-bli.	octet 4*
1 ext	0 access id.	0	rate adaption		signalling access protocol			octet 5*
0/1 ext	0 layer 1 id.	1	User information layer 1 protocol				sync/ async	octet 6*
0/1 ext	numb. stop bits	nego-tia-tion	numb. data bits	user rate				octet 6a*
0/1 ext	intermed. rate		NIC on TX	NIC on RX	Parity			octet 6b*
1 ext	connection element		modem type					octet 6c*
1 ext	0 layer 2 id.	1	User information layer 2 protocol					octet 7*

Fig. 11.16 *Bearer Capability Information Element (BCIE) options*

For mobile originated calls, the user must set the BCIE parameters through the mobile telephone keypad or through the 'R' interface from a DTE.

For mobile terminated calls, setting the BCIE is more complex, particularly in the case where the call originates in the PSTN, since there is no mechanism by which the BCIE can be set by the calling party unless specific functionality is embedded into the IWF.

One proposed method to solve the problem of mobile terminated calls will be to use multiple mobile numbers (i.e. assign a mobile with an alternative telephone number for data). When a mobile's data number is called, the mobile will respond with its own BCIE values. If the mobile's BCIE values are compatible with those assigned by the IWF; then the call will proceed, otherwise the call will be terminated. It will be difficult in the case of the PSTN to convey the clear cause to the calling party.

It is probable that mobile terminated data calls will be for specific and dedicated applications, as indeed is the case in the TACS network, and therefore the BCIE can be preset for a particular application.

11.3.5 The Inter-Working Function (IWF)

The GSM specifications do not define the design of the IWF. However, the interworking between the public land mobile network (PLMN) and the fixed network environments are specified by GSM in so far as the mapping of call establishment, call progress and call clearing procedures are concerned.

11.3.5.1 PSTN interworking

Interworking between the PLMN and the PSTN is by far the most complex and will probably be the most commonly used service in the UK.

In most IWFs, a modem pool will be provided. The modems in the modem pool have to be capable of interworking with a multitude of modems in use in the fixed network; e.g. V21 (300 bit/s full duplex), V22 (1200 bit/s full duplex), V23 (1200/75 bit/s asymmetric full duplex), V22bis (2400 bit/s full duplex), V32 (9600 bit/s full duplex). The latter two modems also invariably offer V42 error correction.

At the time of call establishment, the modem in the IWF has no knowledge of the type of modem in the PSTN and so, for mobile originated calls, the calling modem must automatically configure itself to be compatible with the answering modem. Whilst this is standard practice in PSTN to PSTN connections, there is an added complication for GSM. The data rates between the mobile and the IWF and other BCIE parameters are non-negotiable for mobile originated calls. Therefore, if the rate defined in the BCIE is 2400 bit/s and the modem trains to V22 (1200 bit/s), then flow control for user data will be required, which may not necessarily have been specified in the BCIE. In certain cases the call will be terminated, in other cases the call will be permitted to proceed with the risk of loss of user data due to buffer overflow.

11.3.5.2 Packet Switched Data Network (PSDN) interworking

A number of possibilities are provided to enable interworking with PSDNs:

(a)   An asynchronous DTE in the mobile via a PAD in the IWF.

(b)   An asynchronous DTE in the mobile using dial up access using the modem pool as described earlier.

(c)   An X25 DTE in the mobile via direct connection in the IWF to the PSDN

In the case where mobile X25 terminals are to be used, layer 2 of X25 must be terminated in the mobile data adaptor and regenerated in the IWF. This adds a degree of complexity to the data adaptors in the mobile and in the IWF.

11.3.5.3 ISDN interworking

The GSM specifications allow for interworking with the ISDN.

It is unfortunate that at present the data rate across the air interface limits the ISDN rate to 16 kbit/s. This means that for the immediate future only those fundamental services in the fixed network ISDN which function within the limited data rate will be capable of being supported in the mobile ISDN environment.

11.3.6 The facsimile teleservice

The characteristics of group 3 facsimile machines and their operational limitations in a TACS cellular environment were described earlier.

In GSM, there are some additional technical and operational considerations which complicate the attachment of a conventional group 3 facsimile machine to the mobile telephone.

The speech coders used in the mobile and the GSM network emulate a 3.1 kHz PSTN path but cannot cope with the modulation used in the facsimile message phase. It is therefore necessary in both the mobile and in the IWF to incorporate fax adaptors as shown in Fig. 11.17.

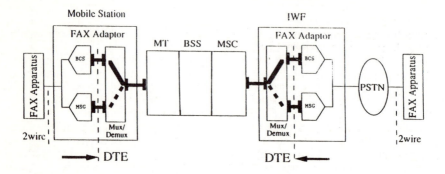

Fig. 11.17 *Fax adaptor*

The fax adaptors have to terminate the facsimile machine modem functionality and elements of the T30 protocol and to signal to each other when to change between the T30 binary control phase (BCS) and the message phase (MSG).

The mobile fax adaptor also has to incorporate some of the properties of a PABX in order to generate ringing current for incoming calls and to convert DTMF or loop disconnect dialling from the facsimile machine into V25 bis auto calling procedures for outgoing calls.

Two Facsimile services are provided by GSM:

- The transparent facsimile service

- The non-transparent facsimile service

The transparent facsimile service allows the T30 protocol to overlay the forward error corrected (FEC) and interleaved transparent bearer described earlier. The transparent facsimile service emulates as far as possible the performance of a PSTN circuit with the exception that it has a fixed delay of several hundreds of milliseconds. The design of the fax adaptor has to take this delay into account since such delays would otherwise adversely affect the CCITT T30 control phase ARQ mechanism.

The non-transparent facsimile service overlays the CCITT T30 protocol onto RLP. In so doing, a degree of complexity is introduced into the fax adaptor design since the T30 control phase protocol is unable to cope with variable and sometimes large delays which may be caused by the ARQ mechanism of RLP. Facsimile machines cannot be flow controlled for very long and therefore buffering is provided for the message phase. It is also necessary for the fax adaptor to absorb and discard some of the control phase commands in order to overcome the effect of the long delays.

The non-transparent facsimile service will in all probability be prone to premature call termination, whereas the transparent facsimile service will from time to time have the performance characteristics of a poor quality PSTN connection.

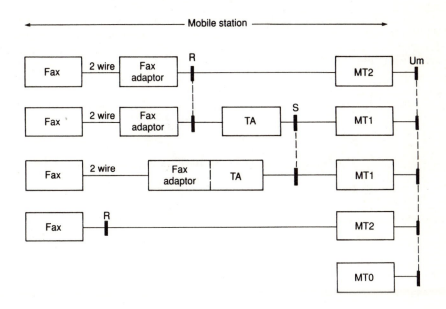

Fig. 11.18 *Group 3 fax connection to GSM*

Fig. 11.18 illustrates a number of ways in which a group 3 facsimile machine may be connected to a GSM telephone.

Operational problems will exist for some facsimile machines having a telephone handset, since that handset cannot be used for speech but may have to be used to initiate or answer a facsimile call.

At the time of writing there is little indication that neither a stand alone fax adaptor or a fax machine with an 'R' interface are imminent. It is possible that PC facsimile card manufacturers may be a little more adventurous.

11.3.7 The short message service

The short message service allows alphanumeric text messages of up to 160 characters to be sent to and from a GSM telephone via a short message Service Centre (SC).

A key feature of the short message service is that short messages may be sent irrespective of whether there is a speech call in progress as they are not sent on the traffic channel.

The point of origin and destination of a short message is known as a Short Message Entity (SME).

An SME may be a GSM mobile telephone or a device outside the PLMN such as a terminal on an X25 network.

11.3.7.1 The Service Centre

The prime function of the SC is to receive short messages from an SME, store them and then attempt to deliver them to the receiving SME.

Once the SC has received the short message from the originating SME, there is no requirement to maintain the communication link to that originating SME. The short message service is therefore essentially a store and forward service.

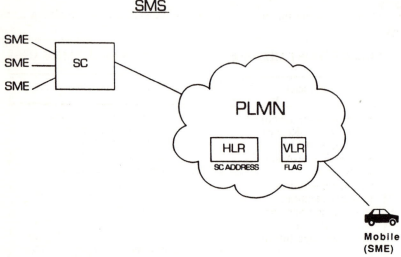

Fig. 11.19 *Key functions of short message service*

11.3.7.2 Outline of operation

Fig. 11.19 shows the key functions of the short message service.

When an SC wishes to deliver a short message to a mobile, it must first interrogate the Home Location Register (HLR) to obtain routing information for the short message upon receipt of which a delivery attempt may be made.

If the delivery of the short message is successful then the mobile will acknowledge its receipt.

It is possible for the mobile to be unavailable because it is perhaps switched off or is temporarily out of coverage. In such a case, the delivery attempt will fail and the SC will retain the short message until the network informs the SC that the mobile is available. At the instant the delivery attempt fails, a flag is set for that mobile in the Visited Location Register (VLR) and a message waiting condition holding the SC address is set in the HLR. When the mobile becomes available, the VLR and HLR are aware of a pending short message and the appropriate SC is notified accordingly. The SC will then attempt another delivery. This feature is known as the short message 'alert'.

The GSM specification does not cover the design of the SC or the means by which it is connected to the PLMN. Neither does it specify how an SME in the fixed network communicates with the SC. It is the latter point which has recently received some attention by those wishing to promote the use of the short message service and discussions are in progress to examine standardisation of access protocols.

11.3.7.3 Features of the short message service

A number of features are provided by the GSM short message service specification. Some of these are described below.

11.3.7.4 Submit short message

This term is used to describe the operation by an SME when it sends a short message to an SC.

The submit operation includes a number of parameters other than the text message itself; probably the most important being the destination address (i.e. the final address to which the short message is to be sent). A status report may also be requested so that the mobile may be told, for example, when a delivery to a destination address has been successful.

Fig. 11.20 shows the full list of parameters associated with submitting a short message.

SUBMIT SHORT MESSAGE   DELIVER SHORT MESSAGE

Parameter	Parameter
Short-Message-Identifier Status-Report-Request Destination Address Service-Centre-Address Reply-Path Originating-Address Protocol-Identifier Data-Coding-Scheme Validity-Period Short-Message-Information	Short-Message-Identifier Status-Report-Indication Destination Address Originating-Address Service-Centre-Address Reply-Path Protocol-Identifier Data-Coding-Scheme More-Messages-to-Send Service-Centre-Time-Stamp Priority Short-Message-Information

Fig. 11.20 *Short message parameters*

11.3.7.5 Deliver short message

This term is used to describe the operation by the SC when it sends a short message to an SME.

The deliver operation includes a number of parameters other than the text message itself; probably the most important being:

- The originating address (if supplied by the sending SME)

- The time and date that the short message was sent into the SC

- The SC address

Fig. 11.20 shows the full list of parameters associated with delivering a short message.

11.3.7.6 Operation commands

The mobile may carry out a number of other operations on a short message which it has previously submitted (provided it is still held in the SC):

- Enquire on its status

- Delete a short message

- Cancel a previous request for a status report.

11.3.7.7 Replace short message

Because of storage limitations for short messages in the mobile, the SC is able to indicate to the mobile that a short message is a replacement short message for one which has been previously delivered. In this way, messages conveying basically the same information may be updated without incurring additional memory usage in the mobile.

11.3.7.8 Display immediate

The SC is able to indicate to a mobile that a short message is for 'immediate display' on receipt rather than the normal mechanism which mobiles will probably adopt which will be to display the message at the user's request. Such a feature is useful where text information is to be conveyed from one user to another whilst a speech call is in progress between them.

11.3.7.9 Ignore short message

An SC may indicate to a mobile that the mobile may acknowledge receipt of a short message but may ignore its contents. Such a feature is useful if the alert feature described earlier is to be used merely to ascertain whether or not a mobile is available.

11.3.7.10 Memory capacity exceeded

This feature allows a mobile (whose memory capacity for short messages was full and now has available space) to notify the SC so that any short messages awaiting delivery to that mobile may now be delivered.

11.3.7.11 The short message mobile telephone

A mobile telephone having short message service capability has to be capable of receiving, storing, displaying and sending short messages.

Short messages received by the mobile from the SC will normally be stored in the Subscriber Identity Module (SIM) card. The SIM card is a feature of all GSM phones and allows a user to remove their personal identity (i.e. their phone number) from the telephone, together with any short messages that they may have received. A SIM currently has a typical capacity for 5 short messages. Any other short messages may be retained in the telephone's own internal memory but access to them must be denied if another user inserts their own SIM into the telephone.

The display area on a GSM telephone is restricted typically to 4 lines of 12 characters. Mobile manufacturers are aware of the need for 'ease of use', particularly for short messages received which exceed the normal display area. A number of interesting display and scrolling methods are beginning to emerge.

Sending short messages from a mobile telephone is a difficult process since a ten digit key pad is all that is provided for the user to create alphanumeric text. Additionally, it will be necessary for either the mobile or the user to insert other parameters associated with submitting a short message as listed earlier. The average user will understand very little about such parameters apart from the destination address and, possibly, the service centre address and so, in all probability, default values will be set by the telephone itself, allowing editing if desired.

It is possible for a terminal to be attached to the 'R' interface, which overcomes the limitations of the key pad and display on the telephone. The terminals may be either intelligent terminals (e.g. a PC) or unintelligent terminals (e.g. a dumb terminal or a PC with terminal emulation software). In the case of unintelligent terminals, it is only possible to enter the destination address and the text message itself. Other parameters are preset in the mobile telephone and if changes to those parameters are required, then it must be done via the mobile telephone keypad. Where intelligent terminals are used, any parameter may be set by the terminal. The penalty for this latter case is that the interface protocol is more complex than that used for unintelligent terminals.

It is a widely held view amongst a number of cellular network operators that the short message service is second in order of importance to all GSM cellular services; the first being speech, of course. The UK has a well established and highly developed TACS cellular network and, at present, the short message service is seen by operators as a significant difference between GSM and TACS phones.

The key interest from most PLMN operators is the integration of the short message service with voice messaging and for applications where the conveyance of text information is more efficient and accurate than voice translation.

11.3.8 The cell broadcast service

The cell broadcast service enables a GSM mobile telephone to receive and display information which is broadcast on a regular basis in a similar way to the UK television Teletext/ Ceefax services.

The information will be broadcast from GSM base stations and the mobile user will be able to select the information required through the telephone keypad. There is no mobile user interaction with the base station or any other part of the PLMN for this service.

As with the TV Teletext/Ceefax services, the refresh rate for updated information is slow and gets worse the more topics there are to be broadcast.

GSM has standardised the 'page' numbering for some of the more common topics such as weather reports and road traffic information.

The business case for freely broadcasting information of any value is difficult to justify. Revenue will undoubtedly come from the need for the user to take some other fee-paying course of action for more detailed information (e.g. a premium rated telephone call). Such a precedence already exists for some of the TV Teletext/Ceefax services.

11.3.9 References and further reading

1. Balston, D.M., 'Pan European Cellular Radio'. *IEE Electronic and Communications Engineering Journal,* 1989, **1** (1)

   GSM Specifications as follows:
   - 07.01 General TA functions for mobile stations
   - 07.02 TA Functions for the asynchronous services
   - 07.03 TA functions for synchronous services
   - 04.21 Rate adaptation at the MS BSS interface
   - 03.10 GSM PLMN connection types
   - 04.22 RLP
   - 03.40 Short Message service
   - 03.41 Cell Broadcast service
   - 03.45 Transparent Facsimile Service
   - 03.46 Non Transparent Facsimile Service
   - 09.xx PLMN/fixed network Interworking

11.4 Summary of data services over cellular radio

The growth of cellular radio has enabled organisations to extend their everyday office-based operations to the mobile environment. Inevitably, this has included data and facsimile.

The problem with 'data' (excluding facsimile) is that there are too many standards to promote its widespread adoption even in the fixed networks. Although there has been some growth in electronic mail, its use is generally confined within a particular organisation. It is not surprising therefore that the use of 'data' in cellular environments is small.

Other than speech, facsimile is the most commonly used method of communications worldwide. The reason for the success of facsimile is its ease of use, primarily because it is standardised as an end to end service. The cellular environment allows the use of facsimile machines at remote locations not served by telephone land lines. However, the performance of facsimile machines under certain conditions is unpredictable, both in the TACS and GSM environments, and in neither case is the attachment to a cellular telephone simple or cheap.

However, the investment in facsimile is now so large that it seems unlikely that even the much needed promise of the ISDN to provide an alternative for national and international communications will make a significant penetration in the foreseeable future.

Until there is an incentive to change attitudes towards data communications in the fixed network, it is unlikely that we shall see any significant growth in data communications in the cellular network. The use of data over cellular radio is likely to remain, therefore, confined to the specific needs of organisations and niche markets for a long time to come.

Chapter 12

Broadcast systems with data capability

John P.Chambers

12.1   Introduction

Broadcast services began with sound radio in the 1920s, followed by television in the 1930s. Since the early 1970s there has been an increasing interest in the use of data broadcasting, to quote the BBC's Royal Charter, 'as a means of disseminating information, education and entertainment'.

Some data broadcasting services are intended to display information on suitably-equipped television or radio receivers. The teletext system allows pages of text to appear on the television screen, either instead of the picture, superimposed on the picture, or inserted in the picture as a subtitle or newsflash. The RDS (Radio Data System) system allows station identification labels and short messages to appear on an alphanumeric display on a radio receiver. The new DAB (Digital Audio Broadcasting) system will support a variety of information accompanying the sound programme, such as that provided with the new (DCC) Digital Compact Cassette, together with details of alternative and forthcoming programmes.

Other data broadcasting services are intended to provide information for devices other than a domestic receiver. Teletext has been used to send software and data for home microcomputers, and commercial data broadcasting services such as Datacast (BBC Registered Trade Mark) used teletext as the data carrier. The signalling capacity of radio-data on the 198 kHz Radio 4 UK service is used to carry the Radio Teleswitching which controls domestic and industrial electricity meters offering a time-variant tariff.

Until recently, sound and vision broadcasting had always used analogue techniques and data capacity had been added many years later as an extra facility. We now have hybrid television systems in which the sound is broadcast in a digitally-coded form, whilst the vision remains as an analogue signal. NICAM 728 stereo for television and the MAC/packet family are of this type. DAB will be the first all-digital broadcasting system and it will provide a very flexible and rugged multiplex which can carry six or more sound services together with related and unrelated data.

This chapter describes the technical basis of the main broadcasting systems in which a data-carrying mechanism has been defined. Many of the principles used in other data communications networks will be found in these examples, together with some features peculiar to broadcasting. Many satellite broadcasting services provide a range of additional subcarriers which can carry information in almost any format, analogue or digital, for any purpose. These peripheral systems are ignored here, as they are not part of the broadcasting service.

The systems to be discussed in detail are summarised in Table 12.1.

System	Carrier	Type	Modulation	Peak bit/s	Mean bit/s	Block length	Useful length	CRC	Additional details
On-screen TV signalling	patches on picture		NRZ brightness	25	25				photo-electric pick-up attached to TV screen
Radio telesoftware	sound channel		FSK	1200	1200	8k (typ)		16	higher bit rates possible using VHF/FM audio channels
LF radio-data	LF radio		bi-phase + PM	25	25	50	32	13	used on 198kHz for "teleswitching" service
RDS	VHF/FM radio	FDM	bi-phase + DSBSCAM	1187.5	1187.5	26	16	10	data differentially encoded to avoid carrier ambiguity
Teletext	television	TDM	NRZ	6.9375M	(1)144k	360	320	16	CRC on block groups (pages)
Datacast	"	"	"	·	(2)54k	360	288	16	variable address length
NICAM 728 -mono sound -no sound	television	FDM	DQPSK	728k	11k 363k 715k	728 728 728	11 715/11 715		Scrambled, bit-interleaved " "
C-MAC/packet D-MAC/packet D2-MAC/packet	television " "	TDM " "	2-4 PSK duobinary duobinary	20.25M 20.25M 10.125M	(3)3.0M (3)3.0M (3)1.5M	751 751 751	728 728 728		Scrambled, bit-interleaved " "
DAB	Multiple carriers	FDM+ TDM	COFDM DQPSK	(3)2.3M	(3)2.3M	50,000 (approx)			Convolutionally encoded time- and frequency-interleaved

1) assuming eight lines per television field
2) assuming three lines per television field
3) this includes as many (digital) sound signals as are required at that time

Table 12.1 Summary of broadcasting systems capable of carrying data

12.2   Aspects of broadcast systems

There are many differences between a broadcast network and a typical data communication network. Existing sound and television networks provide almost complete coverage nationwide with high reliability. So when a data system is added to an existing network it enjoys the benefit of an immediate wide range of potential users. There are thousands of transmitters but tens of millions of receivers. The market size is such that mass production techniques are used in receiver manufacture, and components, notably integrated circuits and display tubes, are developed primarily for this market. Both teletext and DAB were developed on the assumption that dedicated integrated circuits would be produced, and this influenced the design of the systems themselves. Economies of scale have reduced the real cost of receivers steadily over the years and these benefits apply also to receivers for associated data services. Because of the one-to-many nature of broadcasting it is worthwhile to concentrate difficult or expensive operations at the source rather than to divide them between the source and the receiver.

In most parts of the country it is very easy to receive broadcast services, no wires need to be laid and there is no waiting time. DAB will bring the added advantages of a system designed for mobile as well as fixed receivers.

12.2.1   Problems of broadcast systems

New broadcast systems, particularly additions to existing systems, have to be developed with great care. It is most important that they do not interfere with the normal operation of receivers already purchased and in use, even if these are over ten years old. The designer of a new system does not know how many different types of receiver are in use, or how well they are aligned and maintained. In the limit, the only way to prove the compatibility of a new broadcast system is by carefully controlled tests using the public network.

Users of an additional data service expect to be able to receive it without problem if they already enjoy good reception of the associated sound or television service. This too constrains the designer of a new system. The successful introduction of stereo radio and colour television required careful planning over many years.

International agreements cover the allocation of transmitter frequencies and powers, and there are agreed tolerances which allow new services to be planned without interfering with existing services. Whereas telecommunications authorities can provide gateways with agreed interfaces between countries, there are no rigid boundaries to broadcasting. The cross-border and satellite reception of broadcast services makes the work of the European Broadcasting Union (EBU), the European Telecommunications Standards Institute (ETSI) and the International Telecommunications Union (ITU) towards common standards increasingly important.

In most cases, data systems have been added on to existing services where there has been little spare signalling capacity within the channel. In order to make best use of this unique and very limited resource, the overheads such as synchronisation and error control need to be kept very low and special techniques have been developed. The pattern of errors can vary very widely between different users at the same time and there is, of course, no return path available for use in error correction.

The very accessibility of broadcast signals without the need to make a physical connection makes it particularly difficult to ensure the privacy of any messages intended only for a limited audience, such as would be required in a subscription service. Considerable work has been done on developing such one-way access control methods for television and the key management techniques can be applied directly to data broadcasting [1].

Unlike the data service provider, the broadcaster has no control over the manufacture of receivers or the preference of the consumer. This makes it essential that the technical specification of any new service be complete and unambiguous so that it can be interpreted consistently anywhere in the world. Even then there may be problems, and difficult decisions may need to be taken with several thousand sets already sold and in use. Such problems are very much more likely in digital systems as it becomes impossible to check the response to all possible bit pattern combinations and as receiver designers take a pride in providing their own special features by augmenting the decoder software.

12.3 Methods of adding data capability

Most of the systems listed in Table 12.1 have been additions to existing broadcast systems. The exceptions are the MAC/packet family of systems [2], which was conceived as a new television standard incorporating a packet multiplex primarily to carry digital sound but with spare data capacity, and DAB [3] which is the first broadcast system in which all the information is carried in digital form.

12.3.1 Substitution

Perhaps the simplest method of providing data capacity in a sound or television channel is to use all or part of that channel itself to carry data.

The use of a broadcast sound channel to carry data was pioneered in the Netherlands in 1978 in a technical magazine programme called 'Hobbyscoop'. Software for various home computers was broadcast ('radio telesoftware') as a sound signal of the form commonly used to record data and programs on a domestic audio cassette recorder. A highest common factor language 'BASICODE' was devised and software was published for interpreting this on several popular home computers.

As well as being used on the VHF/FM network these transmissions were broadcast using the MF and HF transmissions and they were enjoyed by enthusiasts overseas. In 1984 the BBC broadcast software using this same standard as part of a radio programme called 'Chip Shop'. These radio telesoftware transmissions required no special equipment to receive them; only a radio cassette recorder, or a cable linking the audio output of a radio to a cassette recorder, was needed.

Channel 4 television in the UK carried a programme '4 Computer Buffs' in which data was transmitted by a pulsating patch on the picture using a conventional asynchronous serial data protocol with two brightness levels corresponding to binary zero and one. A photoreceptor was held on the screen by a sucker or an adhesive pad and a small interface board provided a standard logic signal for coupling to a computer.

Clearly both of these substitution techniques intrude into the normal use of the broadcast channel. Although a flashing patch in the corner of the screen is not likely to inconvenience a viewer who happened to find that service the high-level mid-band audio signals of radio telesoftware came as a shock to an unprepared listener. They are unlikely to find regular use in normal programmes although they are well suited to the applications in which they have been used. There are proposals to carry brief periods of in-band data on sound services at programme junctions, using a more efficient and less disturbing signalling method.

12.3.2 Mixed modulation

The LF radio-data service [4] uses phase modulation applied to the carrier of an amplitude-modulated radio transmitter. In order to minimise interference with the normal operation of existing receivers the peak-to-peak phase deviation is restricted to 45° and the bit rate corresponds to frequencies below the audio passband. Because the 198 kHz carrier signal is widely used as a frequency standard, the data to be broadcast is first biphase coded so that every data bit provides a complementary pair of phase changes and there is never any aggregate phase change. Moreover, the data clock is normally obtained by direct division of the carrier frequency so that the modulation is coherent.

12.3.3 Frequency-division multiplex (FDM)

Data to be added to a signal using the FDM technique is first modulated on a subcarrier. In RDS this subcarrier is added to the signal which in turn modulates a VHF/FM radio transmitter. In NICAM 728 the data modulation is added to a video signal which is then transmitted using vestigial sideband amplitude modulation.

12.3.4 FDM on FM sound systems

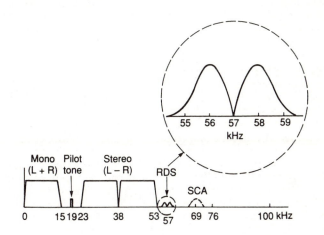

(*a*) Baseband spectrum of VHF/FM sound signal

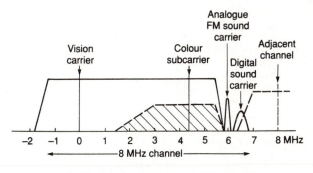

(*b*) IF spectrum of system I television signal

Fig. 12.1 *Data broadcasting using FDM techniques*

Fig. 12.1(a) shows the spectrum of the signal modulating a VHF/FM transmitter. Originally these signals carried only a monophonic sound signal with a bandwidth of 15 kHz. Later the pilot-tone stereo system, which had been designed for compatibility with existing receivers, was introduced. The left-minus-right stereo difference signal is amplitude modulated using double sideband suppressed carrier on a 38 kHz subcarrier. The stereo difference signal, like any other sound signal, is bipolar with a symmetric long-term amplitude distribution function centred on silence. The use of suppressed carrier modulation means that there is no energy present in this part of the spectrum when there is no signal, and this minimises any interference effects. In order to demodulate this signal a 19 kHz pilot tone, obtained by frequency division of the 38 kHz carrier, is added to the signal. This, of course, conveniently appears midway between the monophonic signal and the lower sideband of the difference signal. This frequency is doubled in the decoder to drive the demodulator.

Clearly there must be a limit to this extension of the spectrum of the modulating signal, and CCIR Recommendation 450-1 provides for a maximum carrier frequency of 76 kHz. In practice it is not a sudden limit and it depends on the assumptions made in planning the transmitter network, in particular the spacing of the VHF channels. In Europe this spacing is 100 kHz and in North America it is 200 kHz. So there is potential for including other services at the upper end of the modulating spectrum. But as the transmitter carrier power and peak frequency deviation are also fixed by planning agreement every extra signal could also produce unwanted effects in existing stereo or monophonic radio receivers, due perhaps to intermodulation effects. So any new system must be test rigorously for compatibility.

The VHF/FM radio data system RDS [5] was developed from work done by the Swedish Telecommunications Administration on a paging system. It is based on the use of a carrier at 57 kHz, which is phase-locked to the third harmonic of the pilot tone when present. This means that many of the possible sources of interference become constant or even zero frequency so reducing their subjective annoyance. Advantage in decoder design is gained by locking the data clock to the subcarrier and the data rate of 1187.5 bit/s is the result of division of 57 kHz by 48.

A system developed in Federal Germany to alert road users to broadcast traffic announcements, known as Autofahrer Rundfunk Information (ARI), also uses information on a 57 kHz subcarrier and the RDS system was required to be compatible with ARI. The messages in ARI are carried as single tones of low frequency (125 Hz and below) amplitude modulated on the 57 kHz carrier, so the data modulation used for RDS provides a null at 57 kHz with very little energy to interfere with the sideband of ARI tones. Conversely the ARI sidebands occupy a small portion of the RDS spectrum as shown in the inset to Fig. 12.1(a).

Within the USA a system known as 'storecasting', because of its use to provide background music in shops, but more formally known there as Subsidiary Communications Authorization (SCA), was in use for many years. It used a subcarrier at about 69 kHz which was frequency modulated by the sound signal.

In the USA the Broadcast Television Systems Committee (BTSC) has defined a multichannel television sound (MTS) system which is frequency modulated on a carrier accompanying the vision signal. It uses the pilot-tone stereo system for the main television sound signal, but with the pilot tone corresponding to their television line frequency (H) of 15734 Hz to minimize interference problems between sound and vision. The stereo difference signal is centred on 2H and a Separate Audio Program (SAP) signal is frequency modulated on a 5H subcarrier. It has been proposed that this SAP channel, intended for a second language or audio descriptive service, carry alternatively data at 19.2kbit/s using FSK or BPSK. A further narrow-band FM signal (NPC) on a carrier at 6.5H is also specified. This too could carry data services.

12.3.5 FDM in television systems

The UHF UK television signal (System I) is shown in Fig. 12.1(b). The vision signal has a nominal bandwidth of 5.5 MHz and it is amplitude modulated on the vision carrier with a vestigial lower sideband. Within the vision signal, which is predominantly the brightness (luminance) component, two colour difference signals are carried using suppressed carrier QAM with a subcarrier at about 4.43 MHz. The monophonic analogue sound signal is frequency modulated on a carrier 6 MHz above the vision carrier. The service is planned with a 8 MHz channel spacing so there is little free spectrum (see Fig. 12.1(b)) between the sound carrier and the vestigial lower sideband of an interfering signal on the next higher channel. In continental Europe systems with nominal vision bandwidth 5 MHz, 5.5 MHz sound subcarrier and channel spacing of 7 MHz (System B) or 8 MHz (System G) are in use. An extra FM sound carrier is sometimes used to provide a second language sound service, or the two can be combined together to provide a pair of sound channels for stereophony.

In the UK a system for carrying a 728kbit/s data multiplex on a carrier at 6.552 MHz has been adopted, primarily for carrying to digital sound channels for stereophony or second language applications [6]. Differentially encoded quadrature phase-shift keying (DQPSK) is used and the bandwidth of the resulting signal is about 728 kHz, as indicated in Fig. 12.1(b). Tests have shown that it can be added compatibly to the UK television network. In countries using Systems B and G the same multiplex rate can be carried on a 5.85 MHz carrier with the DQPSK signal filtered to about 510 kHz bandwidth.

This system is known as NICAM 728, from Near-Instantaneous Companded Audio Multiplex and the bit rate. The NIC technique involves quantising the sound signal to 14-bit accuracy and then reducing the accuracy of each 1ms block of samples to at most 10 bits per sample according to one of seven coding ranges. The sample rate is 32 kHz, one parity bit is added per sample, and the coding range is signalled by modifying the parity bits [7], so each sound channel requires 352 kbit/s. Synchronisation and control require 13 kbit/s and 11 kbit/s capacity is available for use to carry additional data, such as an 'audio descriptive service' for the visually impaired viewer. In the event that either sound channel is not in use its 352 kbit/s capacity becomes available as a data service, with neither in use the total useful data capacity is 715 kbit/s.

12.3.6 Time-division multiplex (TDM)

The vision waveform has always contained a significant proportion (25% in system I) of time not devoted to the active picture information. A sound signal has no edges, so there is no need to signal where they are. A vision signal is scanned left to right (lines) and top to bottom (fields) and it is necessary to indicate the boundaries of the lines and fields within the waveform itself. In order to keep the receiver scanning circuits simple, relatively long periods of non-picture (known as blanking) are provided to allow the scanning waveform to be reset and to restart its linear progression. For some of this time early in the blanking periods the waveform is taken to a synchronisation level outside the white-to-black amplitude range, which triggers the scanning processes in the receiver.

The time allowed for line flyback is 12 μs in 64 μs, and that for field flyback is 1.6 ms in 20 ms. These, together with the synchronising pulses and burst of colour subcarrier frequency to lock the colour demodulator, are shown in Fig. 12.2(a) drawn as if the waveform were itself scanned as a picture. Conventional 625-line television scanning uses interlaced fields where each field scan corresponds to only 312 lines. This means that two successive fields produce interleaved scan lines to provide the complete picture.

The 625-line television waveform was designed over 30 years ago and even the oldest receivers still in use have a better field scan flyback time than was originally assumed. So it has become possible to include new information in part of the vertical blanking interval (VBI), the 25 unused lines between active fields. Broadcasters have taken advantage of the VBI for many years to insert test signals to allow the distribution and transmission network to be monitored, and various signals have been added for control and switching purposes. Provided such signals appear like valid vision waveforms, and that they do not come too early in the VBI (where they would interfere with synchronisation), the worst that can happen is that they appear at very top of the displayed picture on a domestic receiver. Many receivers include circuitry to suppress such VBI signals so they cannot appear on the screen even if within the area of the display scans. The teletext system [8] was developed in the early 1970s when it was foreseen that it would soon be practicable to equip domestic television receivers with data acquisition, storage and display circuitry using Large Scale Integration (LSI) for a small additional cost. The original motive was to carry subtitles but it was soon realised that it could provide a full new service, complementary to television and radio. In addition to subtitling and the normal service of pages (such as Ceefax and Oracle) other services not intended for direct display on a domestic receiver can be carried as teletext pages. Teletext has been used to provide a telesoftware service whereby software for domestic and educational use is distributed on specially coded teletext pages and some broadcasters offer subscription data services organised as teletext pages.

Fig. 12.2(a) indicates the position of teletext data lines in the television waveform. When teletext services began in 1974 two lines per field were used. Now about eight lines per field are used and the number can be increased to a maximum of 16 when compatibility with existing receivers allows, and providing the other uses of the VBI, such as test signals, can be accommodated in some way.

When the teletext specification was written certain codes were deliberately reserved to allow the transport mechanism of teletext to be used to carry other services which would not interfere with normal teletext reception. Datacast [9] was such a service.

12.3.7 Dedicated multiplex

Work in the EBU towards a new common European standard for 625-line satellite television led to the MAC/packet family of systems [2]. For reasons of compatibility the PAL and SECAM colour systems carry the colour information within the luminance passband. Although spectral overlap is minimised disturbing interaction between the two ('cross-colour') can arise in the decoder, as it is impossible to separate the two components in a normal picture.

A new system is not bound by compatibility and the colour and luminance components are carried in sequence by time division multiplex in the MAC (Multiplexed Analogue Component) system. Because different bandwidths are required these components are compressed (that is, the waveforms are 'speeded up') by a factor 3:2 for luminance and 3:1 for colour. Both line and field blanking intervals are still provided, but they are all available for use. Indeed, the whole layout of the multiplex can be varied by control data carried in line 625. The basic format for 625/50 television with standard 4:3 aspect ratio is shown in Fig. 12.2(b). Apart from transitions and a reference clamp level the remainder of the line period carries a packet data multiplex. For reasons of compatibility with the D2-MAC system of reduced capacity, the C- and D-MAC system data multiplex is split in two equal parts. Each carries 99 bits per line and corresponding contributions from successive lines are joined to form a data stream which is then divided into fixed length (751-bit) packets starting from a fixed reference point.

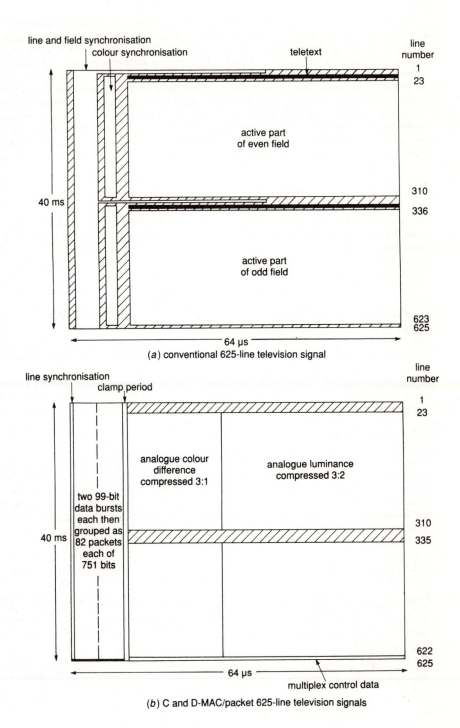

Fig. 12.2 *Data broadcasting using TDM techniques*

The prime purpose of the MAC/packet data multiplex is to carry the sound services required for the programme. As many as eight high quality monophonic sound signals can be carried within the capacity of the C- and D-MAC systems, and lower quality sound options are available, for example for sports commentary in many languages. To the extent that the multiplex is not fully populated by sound packets, which of course must be given priority since sound cannot wait, there is capacity available for any type of data service including, for example, teletext services. It is expected that decoders for the MAC/packet family of systems will provide access to the serial data streams of the packet multiplex for other uses. It seems likely that the rapid advances now being made towards an all-digital television broadcast system will limit the further development of the MAC/packet system.

12.4   Data modulation

A wide variety of the common methods of data modulation are to be found in data broadcasting systems, and the background to the selection of some of these techniques is now given.

12.4.1 Non Return-to-Zero (NRZ)

This very simple technique is used in on-screen signalling where a minimal-cost receiver and decoder are needed. As an asynchronous protocol with start and top bits is used there is no problem over clock recovery.

It is also used in teletext [8] and Datacast where a rugged system with high bit rate is required within the vision bandwidth of 5.5 MHz (5 MHz in continental Europe). The use of odd parity bytes in teletext ensures at most 14 bit periods between data transitions, so simple clock regenerators can be used. More recent decoders use phaselock techniques based on the clock run-in sequence and they can maintain correct clock phase regardless of data transitions in the line. This possibility was anticipated in the teletext specification where the clock tolerance was set at one centicycle per line (25 parts per million). The Datacast specification allows the possibility of totally transparent operation with no restriction on long strings of zeros or ones.

The instantaneous bit rate of teletext is 6.9375 Mbit/s, 444 times line frequency. The recommended pulse spectrum [10] is a 70% raised-cosine roll-off, which is skew symmetric about half the bit rate and reaches zero within the passband. Logic zero corresponds to television black level, logic to one to 66% of while level. This reduced level minimises interference on poorly aligned television sound demodulators yet still allows good decoding of teletext even when the noise is such that picture quality is unacceptable.

12.4.2 Biphase

LF radio-data uses biphase data to phase modulate the carrier in order to provide equal and opposite disturbances to the carrier phase for every data bit. Because there is no long-term accumulation of phase change the carrier remains useful as a standard frequency source.

This absence of a zero frequency component in the biphase signal also explains its use in the RDS system where it is required to minimise energy around the 57 kHz carrier to provide compatibility with ARI.

12.4.3 Frequency Shift Keying (FSK)

Radio telesoftware used FSK as it was the standard in common use for the domestic interchange of computer data and software using audio cassette recorders. The system using a single cycle of 1200 Hz or two cycles of 2400 Hz to provide binary signalling at 1200 bit/s is low cost, well established and adequately rugged for this purpose.

12.4.4 Phase Shift Keying (PSK)

LF radio-data uses low deviation PSK to convey additional information on a carrier without interfering with the normal amplitude modulated sound service. In order to ensure precise and stable spectrum shaping the biphase waveform for the linear phase modulator is generated by direct digital waveform synthesis.

The suppressed-carrier amplitude modulation of 57 kHz by the biphase RDS data is equivalent to a form of two-phase PSK with a deviation of ±90°.

NICAM 728 uses differentially encoded quadrature phase shift keying (QPSK), also known as four-phase differentially encoded phase shift keying (DPSK). This is four-state phase modulation in which each change of state conveys two data bits. The choice of system is dictated mainly by the need to achieve optimum performance within a very tightly controlled bandwidth, although vestigial sideband binary PSK (VSB2-PSK) [11], which had been considered by the EBU for use as a digital modulation system for satellite systems using a subcarrier, was also thought to be suitable.

The C-MAC/packet system switches between two modulation systems at television line rate. The analogue vision components are carried by frequency modulation and the digital information is carried using 2-4 PSK. A one is signalled by a +90° phase change and zero by a -90° change. This is very similar to FSK and it is possible to use the same frequency discriminator for the analogue components and for the data. However, the 2-4 PSK system was chosen to give the best possible error performance when using the entire satellite channel under adverse carrier-to-noise ratios and a more complex demodulator, using differential or coherent techniques, is necessary for best results.

12.4.5 Duobinary

In the D- and D2-MAC/packet systems the binary data stream is converted into a three-level signal using the duobinary technique, and the resulting signal is frequency modulated along with the MAC signal. Conversion to duobinary form concentrates the spectral energy into the region below half the binary signalling rate. So the data component of the D-system has a spectral content comparable to that of the compressed vision signal and that of the D2-system, where the data rate is halved, is suitable for distribution by cable systems using vestigial sideband amplitude modulation.

12.4.6 Differential encoding

The principle of differential encoding where, for example, an incoming logic 1 causes a change of output state whereas a logic 0 leaves the output state unchanged, can be used to eliminate any ambiguity in the polarity of a binary signalling system. It does, however, cause error extension in that a single bit error in the channel causes errors in two consecutive decoded bits.

Differential encoding is used in RDS to avoid ambiguity in the phase of the 57 kHz signal required to demodulate the data. In a two-bit, four-level form it is used in NICAM 728 to avoid the need for an absolute reference carrier phase to demodulate the signal. It is used prior to duobinary coding in the D- and D2-MAC/packet systems in such a way that a continuous logic 0 input results in a continuous mid-level output whereas a continuous logic 1 input gives a continuous high or a continuous low output.

12.4.7 Bit interleaving

Because there is correlation between successive samples of a digital sound signal it is possible effectively to mask the effect of known isolated errors by interpolation between adjoining sound samples. So the NICAM 728 and MAC/packet systems, whose data broadcasting function is primarily to distribute sound in digital form, re-order the bits before transmission and, in a complementary way, after reception in order that bursts of errors in the channel will be converted to isolated errors in the final data stream.

12.4.8 Scrambling

In systems where the structured spectrum resulting from a repetitive data signal might interfere with other signals, or with other components of the same signal, that data is scrambled by modulo-two addition of a predetermined pseudorandom sequence. Digitised silence is such a repetitive data signal and so both NICAM 728 and the MAC/packet data are scrambled in this way. Scrambling also assists some methods of data clock recovery.

12.4.9 Coded Orthogonal Frequency Division Multiplex (COFDM)

Digital Audio Broadcasting uses a method of modulation which had hitherto only been found in textbooks and very exotic applications. Only a brief description can be given here, enough to whet the appetite.

Fig. 12.3(a) shows five 'carrier' waveforms which, between the vertical markers, have in turn two, three, four, five and six complete cycles. The bottom trace is the sum of the five waveforms. For this example, each carrier is in one of four phases at the markers, corresponding to ±sine or ±cosine. Regardless of their phases, these carriers are orthogonal, that is to say the integral of the product of any carrier with any other carrier across the interval between vertical markers is zero. The integrated product of a carrier with its own reference carrier frequency, in two different phases (preferably in quadrature), can be used to determine its phase. These points remain true for an ensemble of carriers with equi-spaced frequencies, over the interval in which each has a different integer numbers of complete cycles. In the case of DAB, there can be over 1500 such carriers. Of course, they are not decoded by 3000 synchronous demodulators, but by a fast Fourier transform (FFT). It is the availability of this technology, at a suitable processing rate, that has made orthogonal frequency division multiplex (OFDM) possible for mass-produced domestic equipment.

The real magic comes when we consider extending these waveforms back in time. For example, the orthogonality interval might be 1 ms and, in the drawing, we extend 300 µs earlier (corresponding to 90 km of propagation distance). Then any linear combination of echo signals (with path lengths less than 90 km), whatever their amplitudes and phases, will simply modify the amplitude and phase of each carrier of the received signal in a certain way. Note that these 'echoes' may even be identical signals from other transmitters, there can be a 'single frequency network'. When the next bits are sent, 1.3 ms later in this example, the pattern of echoes is unlikely to have changed much, even in a moving vehicle. So a demodulator based on the differential phase between symbols is immune to the effects of multipath propagation. Four

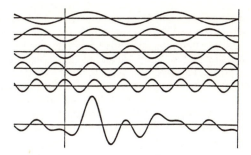

(*a*) Example with five orthogonal carriers and their sum, with a 30% guard interval

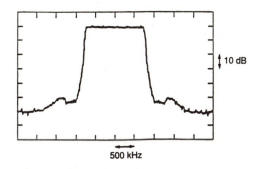

10 dB

500 kHz

(*b*) DAB signal spectrum

Fig. 12.3 *The principles of OFDM*

phases per carrier are used, with constant amplitude, allowing two bits per symbol per carrier. So with about 1500 carriers, allowing for the 30% guard interval in this example, we have a bit rate of about 2.3 Mbit/s. The spectrum of the DAB signal (Fig. 12.3(b)) and its waveform are very similar to those of band-limited noise, and represent a very efficient use of the bandwidth. In order to provide synchronisation and data framing there is a periodic interruption to the signal, followed by fixed initial symbols. This periodicity is linked to the 24 ms frame used in the ISO sound coding system used in DAB.

The other magic in DAB is the 'C' of COFDM, the coding. The data to be sent is convolutionally encoded to give deliberate redundancy, typically adding 75% to the bit rate. The bits from the resulting data streams are then distributed among the carrier frequencies, and delayed by up to 400 ms, so that each particular bit does not depend on a particular carrier at a particular time. Fixed and mobile reception in the presence of multipath propagation is characterised by selective fading, at any particular time and place there may be certain narrow bands of carrier frequencies which cannot be used due to destructive interference. But at a nearby time and/or place the pattern will, in general, be different. The redundant bits coupled with time- and frequency-interleaving completely overcome these problems for most of the time. Even a complete collapse of the signal due, for example, to driving under a bridge, need not affect DAB reception.

12.5   Data packet structure

There have been numerous national and international discussions among broadcasters about the relative merits of fixed data formats and systems supporting packets of variable length. The result is that most systems have a packet-like structure but with fixed-length packets which typically start with address and control information and end with redundant bits to provide an error check. The main examples of variable length packets are in telesoftware, whether by radio or teletext, where the information is sent in blocks typically 200-1000 bytes long and in Datacast where the packet length was adjusted usually to fill but never to exceed the available capacity on a data line. The length and format of the packets depend on the application and some examples are now given.

12.5.1 LF radio-data

Fig. 12.4(a) shows the LF radio-data packet of length 50 bits. At such a slow data rate a long packet would take a significant time to receive and process, even if it only carried a short message. The size was chosen to suit the expected message types and to be a submultiple of one minute in duration, to simplify its use in time-related activities. The blocks are phased so that one immediately precedes the change of minute and this carries a time and date message.

Although a fixed leading bit is provided to assist a decoder in synchronising with the block structure, it is intended that decoders use the 13-bit check word for synchronisation. In effect, the decoder tests all possible block boundaries and latches on to the one which consistently satisfies the redundancy check after every 50-bit block period. Because of the slow data rate this operation can be done by software.

12.5.2 RDS

The RDS format shown in Fig. 12.4(b) is based on packets of only 26 bits, ten of which provide the redundancy check. The packet size was chosen after extensive field measurements of error patterns under adverse reception conditions, such as in a moving vehicle in mountainous terrain. In order to provide a larger message unit these are grouped into blocks of four.

The same implied synchronisation is used as in LF radio-data but with four different ten-bit words added, bit-wise modulo-two, in sequence to the ten-bit check words. The decoder can lock on to this pattern and so achieve group and packet synchronisation.

RDS was developed from a Swedish paging system and codes were provided to allow such a system to co-exist without confusing the decoder.

12.5.3 Teletext and Datacast

The length of the teletext packet is determined by the number of bits which can be signalled within the television active line period of 52 μs. When the display standard of teletext was decided at 40 characters per row of text there was a strong wish to encode information on a one row per line basis. Fig. 12.4(c) shows the structure. Two bytes of 1010... clock run-in sequence were provided to give bit synchronisation followed by an eight-bit framing code to give byte synchronisation.

Addressing was kept to a minimum by using only eight bits, three to indicate one of the eight magazines and five to indicate one of 32 rows, only 24 of which were to be used. These eight address bits were encoded as two 8,4 Hamming coded bytes to give protection against errors. The boundaries between pages of a magazine were marked

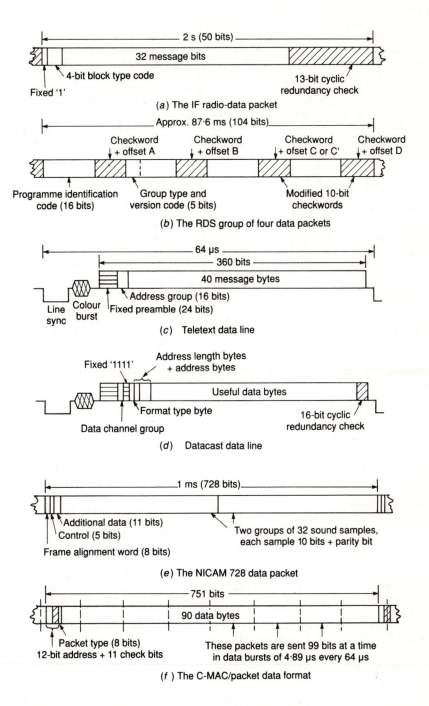

Fig. 12.4 *Broadcast data packet formats*

by page headers (row 0) which, exceptionally, used the next eight codes for further address and control instead of as character codes. So there are 45 bytes (360 bits) in a teletext data line, and in most lines 320 of these bits represent useful data.

Datacast uses the teletext data line format but it uses some of the eight-bit address codes which are ignored by a teletext decoder to provide four independent channel groups. The remaining 40 bytes then become available for redefinition, as shown in Fig. 12.4(d). The first byte specifies the following format in more detail, whether packets are repeated, how the continuity index is sent, and whether the packet is deliberately shortened. The next byte specifies the length of the packet address, which can be 0, 4, 8, 12, 16, 20 or 24 bits. After the address and control bytes come the useful data bytes followed by a 16-bit cyclic redundancy check. Between 28 and 36 bytes of useful data are available per line, depending on the format selected. Unlike teletext, every Datacast line can be checked and interpreted in isolation.

12.5.4 NICAM 728

The NICAM 728 data is gathered into 728-bit packets (frames) of 1 ms duration as shown in Fig. 12.4(e). This fits well with the NIC operation which compands blocks of 1 ms of sound (32 samples). The NICAM 728 packet contains two groups of 32 sound samples, corresponding to a stereo pair of 2 ms of sound from one source, with each sample coded as ten bits plus a parity bit. The range code for each sound block is signalled as a three bit number, each bit being signalled by the parity of nine particular samples being odd or even. The sound accounts for 704 bits of the packet. Eight bits are used for a frame alignment word to give packet synchronisation, and five bits are used for control functions which indicate how the sound channels are used or whether they are available for other data services. The remaining 11 bits per millisecond are available as an additional data channel, and are being used experimentally to carry an audio descriptive service. Although Fig. 12.4(e) indicates that the 64 11-bit sound samples are sent in sequence this part of the packet is bit-interleaved prior to transmission.

12.5.5 MAC/packet

All members of the MAC/packet family use 751-bit packets as indicated in Fig. 12.4(f). The first 23 bits comprise a 23, 12 Golay code which is unusual in being perfectly packed. Twelve message bits are protected by a further 11 check bits and there is a complete correspondence between the 2048 possible check syndromes and the $1 + 23 + 253 + 1771 = 2048$ possible patterns of no, one, two or three errors distributed among the 23 bits. So this code can correct triple errors but, of course, it will falsely "correct" four or more errors. This powerful code is used to send a twelve-bit packed address which, in the case of digital sound, is taken to be a ten-bit packet address associated with a two-bit continuity index to provide indication of packet loss or gain.

The next byte is a packet type indicator, whose function is well defined for sound signals where it allows control packets to be inserted into the series of sound packets. The remaining 90 bytes carry sound samples which, as with NICAM 728, are bit-interleaved.

In normal use the MAC/packet system will have ample spare data capacity, and packet addresses, available for other data services.

12.5.6 DAB

The DAB data is not structured in a packet format although, of course, a packet-based data service can be carried by DAB. The main data is organised into 24 ms data frames which are divisible into several hundred elementary data channels. Fixed numbers of these are required for the various sound coding options and other groups including the residue can be used for other data services. The contents of the DAB multiplex can be varied at any time without disturbing continuing services. The services are 'stacked' so that each uses a 'consecutive' group of elementary channels, which may change when the multiplex is reconfigured.

12.6   Applications

Some of the systems described above were designed for specific applications. The LF radio-data system was primarily intended for the "teleswitching" system for load management and tariff control in the electricity supply industry [12]. The service is very reliable and the frequency used allows reception within buildings even below ground level. The service also offers a source of time and date.

RDS is primarily intended to offer a programme identification service which, together with a list of alternative frequencies which can also be signalled by the system, allows a fixed or mobile receiver to search for a particular programme and to find the best available signal. It also provides the facility to send 64-character messages for display, using an eight-bit coded extended character set to meet the needs of most European countries. A time and date service is also provided, using Modified Julian Date and Coordinated Universal Time to be independent of time zone and calendar convention. As part of a European initiative a traffic message service, using densely-coded pre-defined messages (in seven languages) is being developed. It will use about 100 bit/s of the capacity.

Teletext is perhaps the most widespread and well established of these data broadcasting services. It includes methods of programme delivery control to assist the operation of domestic video recorders. Datacast offers reliable, low cost transparent data channels for a variety of special uses, including the updating of databases and the distribution of specialist financial information to subscribers.

NICAM 728 already provides digital two-channel or stereo sound with television in many countries, and some broadcasters are experimenting with using the extra data capacity for audio descriptive services for the visually impaired viewer.

The MAC/packet system offers great flexibility both as a television system and as a data carrier. Although it is being used in several special cases it seems likely to be superseded as a contender for a new broadcasting standard by a future all-digital system.

DAB is an exciting and totally new method of broadcasting, using modern technology and designed for mobile use. Although its prime application will be to carry digital sound broadcasting, similar techniques will undoubtedly be used for general data transmission.

12.7   References

1.   Wright, Conditional Access Broadcasting: Datacare 2. BBC Engineering Division Research Department Report 1988/10.

2.   Specification of the systems of the MAC/packet family. EBU, Brussels, October 1991, Tech.3258-E (Second edition).

3.   'Radio Broadcast Systems: Digital Audio Broadcasting (DAB) to mobile, portable and fixed receivers'. European Telecommunications Standards Institute (ETSI), January 1994, draft PR ETS 300401.

4.   Wright, L.F. RADIO-DATA: Specification of BBC phase-modulated transmissions on long-wave. BBC Engineering Division Research Department Report 1984/19.

5.   British/European Standard BS/EN 50067:1992. Specification for radio data system (RDS)

6.   Bower, NICAM 728 - digital two-channel sound for terrestrial television. BBC Engineering Division Research Department Report 1990/6.

7.   Chambers, Signalling in Parity: a Brief History. BBC Engineering Division Research Department Report 1985/15.

8.   BBC, IBA, BREMA, September 1976, Broadcast Teletext Specification.

9.   Chambers, BBC Datacast, **EBU Review (Technical),** 1987, (222) pp.80-89

10. Kallaway and Mahadeva, CEEFAX: Optimum transmitted pulse shape. BBC Engineering Division Research Department Report 1977/15

11. Shelswell et al, VSB 2-PSK: A modulation system for digital sound with television. BBC Engineering Division Research Department Report 1986/7

12. Edwardson et al, 'A Radio Teleswitching System for Load Management in the UK'. Fourth International Conference on Metering, Apparatus and Tariffs for Electricity Supply, IEE Conf. Publ. No.217.

Chapter 13

Image networks - A user's view

Ron Bell

13.1 Introduction

With the widespread adoption of personal computers, the end user has become much more aware of the need to communicate between such devices and the problems that such communications introduce. Despite this awareness the user is becoming more and more demanding in terms of movement of computer generated information. As a result any excuse as to why the movement of such information is difficult generally falls on deaf ears. It is therefore important that telecommunications experts take steps to understand fully users' total communications needs.

In order to understand how a user perceives electronic communications it is necessary to appreciate the history of information processing and in turn differentiate between the various types of information. In this latter regard it is only proposed to consider two broad types of information, i.e. *verbal* and *image*, and for the purpose of this chapter image type information will be assumed to consist of all types that are not audible.

13.2 Types of information

Although, with the advent of electronic information processing, we tend to think that information exists in many forms, in fact there are only two which humans can effectively assimilate. These are verbal and *visual*. While humans have the ability to use their other senses, the volume of information that can be interchanged in some other form is limited and as a result, unlike the rest of the animal kingdom, humans have not really developed, for communications purposes, the senses of touch, taste and smell.

While the amount of information which can be interchanged in verbal and visual form is only limited by the capability of human intelligence to absorb it, there are nevertheless two severe restrictions which affect how humans use such information.

13.3 The limitations of verbal and visual information interchange

Regardless of whether information is in verbal or visual form, its effective use is limited by both *time* and *distance factors*, both of which are affected by human capability.

The actual time of receipt of information is often unimportant, providing it is available when it is needed to be acted upon. As a result the time factor is often influenced by the ability to store the information for later recall.

The distance aspect is much more significant as it is limited by the fact that humans can only see and hear over relatively short distances. By considering both of these limitations it becomes much easier to appreciate the user's perception of telecommunications networks.

When telecommunications networks were first being established it was normal to install separate networks for audible and image information. Interestingly the network initially set up to carry image information *(telegraph)* was actually established before the verbal *(telephone)* network. The reason for this was simply because the device necessary to convert verbal information into a form that facilitated its transportation over long distances was invented subsequent to the telegraph (Morse) code sender. However, once the telephone was invented, the essential network to render it of any value developed much more rapidly than its image information counterpart. The reason for this is probably due to the fact that humans learn to communicate verbally, much more quickly and easily than they do using images (pictures and text). Nevertheless, it is also important to remember that, until relatively recently, image type information was very much easier to store for later recall than that in audible form.

A further complication in the perception problem arises nowadays, because it is becoming more and more difficult to differentiate between networks which carry audible as opposed to image type information.

So where has all of this got us? Starting with verbal information we have:

- A device (the telephone) capable of converting sound energy into electrical energy, thus facilitating the task of transporting sound over long distances.

- A ubiquitous network capable of interconnecting telephones on a worldwide basis.

- Further "devices" (people) capable of close association with the simple telephone which are able to interpret and respond to the verbal information, as it is received.

In other words we have an end-to-end network capability with complete functionality.

When we turn to image type information, the picture is not so rosy. First of all the exclusive network which previously dealt with images (text), i.e. the telex network, has, for reasons of economics, been absorbed into the verbal (telephony) network. Unfortunately, while the initial telephone network was ideally suited to carry verbal information and while in the early days it was adequate for limited image type information, as demands for more sophisticated image type information grew, the telex network became more and more inadequate. Anyone who has had the practical experience of trying to transmit information accurately as fast as 300 bit/s over the Public Switched Telephone Network (PSTN) only 30 years ago will appreciate the limitation.

We can therefore summarise where we have got to in historic network terms, as follows. We have a voice network which may, from time to time, suffer from unreliability but, when it is working, serves the end user well, in that it allows any two or more telephones to be interconnected and allows the humans speaking or listening to those telephones to communicate as if they were face to face. They may of course lose some of the value of the information through not also having visual communication, but the adept ones can compensate for this through vocal inflections.

When we consider image communication, all is not well and has been significantly worsened by advancing developments in the production, storage and manipulation of this type of information. As with other types of information it is appropriate to consider the history behind image type information, in order to appreciate the users' perception of its transportation.

13.4 The history of image type information

In order not to dwell on ancient history, suffice it to say that image information has passed through the following phases:

- Drawings on cave walls.

- Drawings, and then symbols representing words, on portable material.

- The invention of the printing press.

At the last point the human race gained the capability to produce copious amounts of image information and store it, almost indefinitely, for future consumption. If it was desired to move the information so generated and stored, initially this could only occur through physical means. This state of affairs prevailed for almost 400 years until in 1840 Wheatstone invented a machine for transmitting letters and figures electrically. This machine was refined over the next 100 years and eventually became the electro-mechanical telex machine. Although this was not to achieve the same penetration as the telephone, the telex network did nevertheless gain worldwide acceptance and the text standard developed for use over it still enjoys today the reputation of being a genuine international standard, albeit with a somewhat limited character set.

Another development did occur during the same timescale and that was the invention of the electronic computer. Interestingly the first mechanical calculating machine, Babbage's Difference Engine, was invented long before the first electrical telegraph message was transmitted. Despite this, it took over 100 years before the first digital computer emerged. Even then these devices were huge and dissipated enormous amounts of electrical energy. As a result it was not until the early 1950s that the first commercial computer appeared.

It became rapidly apparent that to be of any value it would be necessary to be able to communicate with these machines over large distances. The obvious choice of network for doing this was the one that already existed to transport image information, i.e. the telex network. Unfortunately this network was only capable of transmitting at speeds of up to 400 characters per minute, i.e. 53.3 bits per second. Although this was better than nothing, it wasn't much and another device, the modem, was invented for use with the much more ubiquitous telephone network. While this produced a vast improvement in quality and speed, the latter was still limited to no more than 300 bits/s. Fortunately in the early days of commercial computing the effect of this problem only impinged on a relatively small number of specialist users, who tended to understand the technical reasons for the limitations.

As the cost of operating businesses became more significant, it became apparent that by using computers for more and more business applications, operating costs could be stabilised and in some cases reduced. This caused a huge development in large mainframe computers. As a result increasing numbers of non-specialist staff began to use the facilities made available by such machines. Unfortunately for data processing personnel, many of these end users were highly intelligent and began to devise ways of doing their own work more efficiently and quickly by using the processing power of the mainframe; unfortunately because these new applications began to place unscheduled and therefore uncontrolled demands upon mainframes. These demands naturally caused disruption to the applications which didn't need non-specialist input, but which were in fact the main business reasons for adopting mainframes. The disruption consequently had to be brought under control and as a result the data processing departments of most companies became all powerful.

Naturally this caused a significant reaction from the non-specialist users, who objected to having their work controlled by "electronic whizz kids", with little concept of ordinary business practices. As a result of this the distributed processor was developed, followed by the mini- and micro-computer, until we reach the present day when personal computers, possessing capabilities far in excess of many mainframe computers of only 20 years ago, are starting to appear on every desk. These personal computers, or PCs as they are affectionately known, can produce image type information that was undreamed of as little as ten years ago. Regrettably, although we now have an image producing device which is almost as ubiquitous as the telephone, the facilities for that device to communicate in the same manner as the telephone have not developed at the same pace. This leaves us where we are today from the users' perception point of view.

13.5 The user's perception of image type information communication

Individuals that have been part of the private telecommunications business for some years understand the perception that users have of networks. Users do not actually care what technology is employed, their only concern is to be able to establish connections quickly and, once a connection is established, to interchange information as quickly and effectively as they want.

As the pace of business life quickened users did appreciate enhancements to the voice network, such as voice storage for later recall, as well as advanced functions such as call diversion, ring-back-when-free and mobile telephones, all of which are designed to allow interconnection to be achieved more effectively. It is probably true to state that today the end user is generally satisfied with the telephony network and is unlikely to demand many further developments, other than improved mobility.

However, the same situation does not apply as far as image communication is concerned and unless the information technology industry quickly comes up with a solution to users' image communications problems, the users will, just as occurred in the data processing industry, very quickly take matters into their own hands. Before this occurs the IT industry must rapidly appreciate what users are demanding and, equally quickly, put the appropriate capability into place.

13.6 What the users want from image communication

As with many demands it is much easier to define image communications needs than to meet them. With this in mind let me now describe the scenario that end users are facing.

Most business users nowadays have either at their own work space, or nearby, a personal computer which they use for individual business applications. Some of these applications are mundane in that they are merely used to receive image type information and present it in a form that they, or a colleague, can recognise and therefore comprehend. Other applications are more sophisticated in that they allow information to be either received from other sources, or generated within the PC and then processed for representation in another form for subsequent interpretation by someone else. Although the information can nowadays be presented in a number of forms, e.g. text or graphics, and even moving pictures, it is all designed to be presented visually. However, as PCs become even more powerful, it is inevitable that demands to mix both visual and audible information freely will arise. As the networks nowadays are designed to accommodate both types of information in the same digital format, this should, in theory, not present too many difficulties.

On the face of it, what the end user is demanding does not appear too onerous, so let me state it in simplistic terms to emphasise the point. The end user wishes:

- To generate image information in any form (text, graphics, pictures).

- To gain access to stored image information, wherever it happens to reside.

- To transport stored image information from wherever it is stored, into their own PC.

- To transport image information stored or generated in their own PC to some other user.

- To alter and/or process image information within their own PC.

Obviously this list is not exhaustive but I believe it is comprehensive enough to allow the need to be appreciated. In fact what I have described is exactly what the user is capable of doing today, and indeed has been capable of doing for some years, with verbal information. In other words what is first needed of an *image communications network* is the same interconnectability that is offered by the worldwide telephony network. Naturally such capability can be achieved by using the telephony network, but unfortunately the current largely analogue technology network has neither the capacity, nor the quality, to transport the vast amount of image information that will be generated by modern processors.

Equally unfortunately, neither will the immediate future telecommunications network, which possesses the required degree of ubiquity. I refer of course to the *Integrated Services Digital Network (ISDN)*. Although this represents a considerable improvement in bandwidth terms, it is likely to be inadequate for transporting the image type files which will be generated in the foreseeable future. One can gain a good insight into the problem when one considers the file sizes of some applications which are already in use today. For example a Computer Aided Design (CAD) file for an automotive engine can be in excess of 100 Megabytes. Likewise a file containing digitised x-ray information may be up to 12 Megabytes.

As will be appreciated when applications which require such magnitudes of information to be moved, virtually all of today's networks are inadequate, even those with a limited geographic extent such as local area networks. All of this leads to the question of what is needed to meet the users' needs for the future.

13.7   Future image information network needs

Although I have used the word future, in reality many of the demands already exist. Let me give two practical examples, the first the automotive industry and particularly the Ford Motor Company in Europe.

When Ford of Europe design a new car this involves two design centres, Cologne in Germany and Basildon in the UK. Although in the past this presented difficulties involving the movement of many thousands of paper drawings, nowadays with the use of CAD systems, it has become very much easier. Because only two centres are involved, the telecommunications problems are relatively easily resolved. But it is the next two stages in the car production and retail process which present the biggest problems. Although Ford undertake the initial design themselves, once the design moves into manufacturing, the Ford contribution becomes less as their suppliers are expected to undertake the design of individual components themselves. These components are then brought together for final assembly by Ford.

Using this method means that the initial CAD designs must be shared with the Ford suppliers and, as is often the case, when a new design moves from basic design into manufacturing, the initial design has to undergo changes to surmount specific manufacturing difficulties. This effectively means that CAD files, some of which are many tens of megabytes in size, have to be frequently transported to 400+ suppliers around Europe. While the problem could be resolved by means of a large fully interconnected private network, because of today's international bandwidth tariffs, this would be prohibitively expensive. Ford plan to surmount the problem by using ISDN, but of course the bandwidth capability of the ISDN is less than adequate and some compromise on file movement speed may have to be accepted.

The next stage in the process, i.e. when the new car is being sold and maintained, presents even bigger problems. This is because all of the 4,000+ dealerships around Europe which retail Ford vehicles are franchised, i.e. not owned, even though they are closely monitored, by Ford. To be successful these dealerships must be kept up to date with glossy sales literature, service manuals etc. and because of the sophistication of modern vehicles, often have to use problem diagnostics which only reside at Ford premises. All of this effectively means that every dealer needs a high quality, relatively wide bandwidth link back to Ford. Again the need could be met from a functionality point of view by an extensive private network, but the only economic solution for the foreseeable (at least 5 years) future, is the 64 Kbits/s per channel ISDN.

Another practical example can be found in the healthcare industry. As demands on this service, through an growing population, with a longer life expectancy, increase, ways must be found to render it more efficient, but without loss of capability. The net result is that many services, such as x-ray processing, or patient monitoring, will have to be centralised. To achieve this will almost certainly require a network whose bandwidth is far in excess of any existing today. Although, in a similar manner to the automotive industry, the ISDN will be used, this can only be viewed as a temporary measure and something very much better will be needed within a 3 or 4 year timeframe.

13.8   The future information network

You will note that I have now dropped the word image and merely refer to information. The reason for this is that in the very near future the demand for moving information will be universal, i.e. it won't be voice or image, but a totally integrated need. To give it the latest name it will be *"multi-media"*. This implies that the EC directive, which effectively permits national monopolies in public voice telephony networking to be upheld until at least 1999, and in some countries even longer, is going to impose unnecessary restrictions.

On the other hand it is uncertain how any public network operator is going to be able to distinguish between voice and image communications and therefore it is equally uncertain how the EC directive can be upheld practically.

That aspect aside, what is needed for the future is a network that is analogous to an immense, incredibly transparent piece of glass. This "piece of glass" should be capable of accepting information in any format and in any quantity, at any point on its periphery and transporting the information to any other peripheral point, or points, totally unchanged and with a minimum of delay. The points to which the information will be transported will be decided by the user depositing the information. The only essential intelligence the network will need is routing and transportation capability. Although other intelligence may be required, this can be provided at the network periphery, perhaps using add-on techniques. While this latter type of intelligence does not need to be an actual part of the network, it may prove more economic to provide it in that manner.

Obviously the "piece of glass" theory will take care of the transportation problem, but it will not solve the problem of incompatible systems. While standardisation will go a long way to solving this problem, such techniques are liable to stifle innovation. Consequently a more practical solution may appear through the provision of value-added network services, specifically to undertake compatibility functions.

13.9   Conclusion

The reader will by now have gathered that, from a user's perspective, I am not particularly enamoured with the *image transportation services* that are being offered by our national telecommunications authorities. It seems to me that they are much more concerned at locking users in to their services in order to perpetuate their revenue earning potential, than attempting to assess what users actually need to meet future business requirements. It is however unfair to single out telecommunications authorities for these problems, I am equally convinced that the computer industry must take the blame for some problems. For instance the latter has not been particularly foresighted in communications terms and now needs to concentrate on developing systems that are either compatible with each other, or at least have a standard interface.

What is actually needed urgently is a prolonged period of co-operation between the computing and telecommunications industries to develop real business solutions for the future. To achieve this both of these industries need help from academic establishments to produce some real experts in *system Iintegration*, of which the world is so desperately short.

Equally all participants, i.e. system designers and integrators, in conjunction with end users, must form partnerships to ensure that all future developments are mutually beneficial and will result in shared advancement.

Index